Earth Science

James W. Fatherree

Copyright © 2016 by James W. Fatherree

ISBN 978-0-9786194-3-5

All Rights Reserved

No part of this book may be reproduced, transmitted, or stored in any form (electronically, mechanically, or otherwise) including, but not limited to, photocopying, Internet publishing, storage retrieval, and reprinting without explicit written permission from the author.

All images are by James W. Fatherree, are licensed under Creative Commons, or are in the public domain. All modifications to images by James W. Fatherree.

Published by:
Liquid Medium Publications
Tampa, Florida

Earth Science

Table of Contents

Keys to Success	1

Unit 1
1.1 The Universe, Our Solar System, and the Earth	5
1.2 The Earth's Atmosphere, Weather, and Climate	7
1.3 Humidity, Clouds, and Precipitation	12
1.4 Air Pressure and Wind	22
1.5 Air Masses and Weather Fronts	26
1.6 Severe Weather	29

Unit 2
2.1 Minerals and Rocks	35
2.2 Weathering, Erosion, and Mass Wasting	44
2.3 Surface Water and Groundwater	51
2.4 Glaciers	60
2.5 Deserts	67

Unit 3
3.1 The Earth's Structure and Plate Tectonics	73
3.2 Earthquakes	82
3.3 Volcanoes and Plutons	87
3.4 Mountains	102

Unit 4
4.1 Seawater, Ocean Basins, and the Seafloor	106
4.2 Ocean Circulation and the Tides	112
4.3 Coasts and Shoreline Features	118
4.4 Geologic Time	123
4.5 A Brief History of Life	128

Sample Exam Questions and Answers	134
Image Credits	137

Keys to Success:

Get the material:
To start, come to class. Good attendance is critical to success, as there are many things covered in class that need explanation if you expect to understand them. Oftentimes, there are things covered in a class that you'll need to know and understand later in the course, too. Think of it like a math course. If you miss a few days in the beginning, you're likely to have a difficult time for the rest of the course.

It's not always easy, but try to get some sleep, too. You need to stay awake in class and pay attention if you expect to do well. Drink some coffee if needed. You should also try to eat something before class, even if it's just a little snack. An empty, grumbling stomach can be very annoying and distracting, and everyone performs better when they're not hungry.

Speaking of distractions... Do not talk when any instructor is talking. You need to be paying attention, and talking not only means that you're not hearing what the instructor is saying, but also means that anyone else that can hear you talking will probably be distracted. Even if you don't care about something an instructor is covering, that doesn't mean the other students around you feel the same way. So, show some respect for your instructor and your classmates and keep quiet when you should be quiet.

Next, is note-taking. Be sure to do a good job of getting everything covered in class, unless specifically told that you don't need to. Whether it's on a Powerpoint slide or written/drawn on the board, you need to know it unless told otherwise. If you have trouble keeping up, or have a hard time writing and simultaneously paying attention to what's being covered, get help before or after class. All instructors have specified office hours for helping students.

Learn the material:
When students that perform poorly on exams are asked about how they studied, they almost always have the same two answers. They didn't study at all, or they just read over their notes/through the textbook a couple or a few times. While that may work in some classes, or for students with exceptional memories, for the most part it's the least effective way to learn large amounts of information.

Also, students frequently say they really studied and knew the material, but then couldn't remember it when it was exam time. However, this means they didn't really *know* the material. In essentially every case, these same students say they studied by reading and re-reading the material, and again, that's the problem. When most students read over the material the first time, they may go slowly and think everything is soaking in. However, on the second or third time through, students tend to read much faster, skimming over much of the information because it looks familiar. They fool themselves. They see something on the page that they know they've been over before, so they wrongly get the impression that they know it. However, recognizing the information is by no means the same thing as knowing it.

This is why so many students honestly think they knew the material before an exam but then "blanked out" or suffered from "test anxiety". They didn't blank out, rather they never really knew the material in the first place. And for that matter, research has shown that test anxiety is typically just a student's subconscious mind telling their conscious mind that they don't really know the material they're about to be tested on.

Many students also study by re-writing all of the information they're responsible for. This is far superior to just re-reading material, and can certainly be helpful at times. This is especially so when it comes to things like spelling. If you have trouble remembering how a word is spelled, repetitive writing can be very effective. And in case you're wondering, yes, spelling things correctly is very important, especially in a science class.

Still, for the most part, the best way to learn much of the material is to make and properly use flash/study cards. They work, if you will, and countless students that use them have stated so. This is especially true in any course with lots of new words/terms, like Earth Science.

Our brains are generally better at remembering lots of small pieces of information rather than fewer but larger pieces of information. So, making study cards is a good way to break things into discrete units of information that are relatively small in size. When making study cards, you're also re-reading the material, and you're re-writing it, too. In fact, many students are quite surprised at how much they learn by simply making the cards – before actually using them to study. Making the most effective study cards will improve performance, too. For example, let's say you're shown the PowerPoint slide below in lecture:

> **The Three Rock Types:**
>
> Igneous - formed by the cooling and solidification of magma or lava.
>
> Sedimentary - formed by the lithification or precipitation of the weathered products of pre-existing rocks.
>
> Metamorphic - formed by the alteration of pre-existing rocks by extreme heat and pressure.

You might make a card that looks just like the slide. However, that's not the best way to make cards, as the information can easily be broken into smaller (and easier to remember) units. Instead, you should make four cards for the slide, each having a smaller amount of information on it. Like this:

The three rock types:	*Igneous* *Sedimentary* *Metamorphic*
Igneous rocks:	*Formed by the cooling and solidification of magma or lava.*

(front of cards)   (back of cards)

Then make two more cards, with a third card for sedimentary rocks, and a fourth card for metamorphic rocks. Get the idea? This way you've got the three rock types listed on one card, and the definitions of each of the three rock types on their own cards. Doing it this way also means that you need to actually *think about the information* in order to choose the best way to make the cards. Thinking about it, rather than just mindlessly copying it, helps you remember it.

So, that's the basic idea behind making study cards, but there's more to say about them. First of all, you should try to make your cards every week, or even better, after each class meeting. Making them each week while the information is still relatively fresh in your mind is far better than trying to make a big stack of cards a few days before an exam. Making them as the course moves along also means you can use them as the course moves along.

When it's time to make some cards, you should also have the attitude that you're about to study for your exam. Turn off the TV, turn off the radio/mp3 player, turn off your phone, etc. and focus on the material as you make your cards. Again, you can potentially learn a lot of information just by making the cards, but probably not so much if you're distracted and just working like a human copying machine.

When you have some cards made, again, keep in mind that smaller units of information are easier to remember than larger ones. Take 5 or maybe 10 cards and learn them front and back. Read them as many times as you think you need to in order to remember everything on them. Then get 5 or 10 more. Then 5 or 10 more, and so on, until you've worked through all of them.

In order to check your progress, the next step is quizzing. You can quiz yourself, but it's far better to give your cards to someone else, even if they're not a student, and let them quiz you. For example, they could ask you to define a term, and you should define it. Or, conversely, they could give the definition of a term, and you should give the term. For many words it's also best to spell it out loud.

When doing this, make two stacks of cards. One for cards you knew (things you got right when quizzed) and a separate stack for the cards you didn't know. Now there's no fooling yourself. You either get it right, or you don't, and you can see by the size of each stack how well things are coming along.

Being quizzed also helps in another way. Typically, taking notes, making cards, and studying are all about putting information *into your brain*. But, taking an exam is all about getting information *out of your brain*. Thus, our brains are really doing two different things. You can think of being quizzed as a way to practice information recovery, like finding a file or document in your mind and opening it up. It should seem like common sense that only practicing putting information in isn't the best thing to do since your grade is based on how well you pull learned information out.

Anyway, get back to work after being quizzed, but this time take 5 or 10 cards from the "didn't know" stack and focus on those rather than all your cards. There's no need to waste time going over the ones you knew, although you should definitely go through them again at a later time before the exam just to make sure you still remember them. At some point it's also a good idea to mix everything up, like shuffling a deck of playing cards, but with half the deck flipped over backwards. This way you see terms on some of them and definitions on the others, etc.

If you make good cards, and used them as described, you should get the bulk of the information you need to know. You should be able to give a quizzer whatever's on the front and back of each card, list off all the items in a list of things, like the three rock types, etc. But, there's still more.

As mentioned above, if you have problems remembering how to spell something, nothing works better than re-writing the word a few times. So, you may do something like putting a little checkmark on any card with a term on it that you knew, but couldn't spell correctly. That way you can easily pull them out of the stack at some point and work on just the terms you need to work on.

Of course, you'll want to look over any pictures and diagrams you need to know, too. No matter how good or bad an artist you are, re-drawing something is also a good way to remember it, just like re-writing words. Likewise, if there's any math involved with the material, be sure to re-work any examples or problems given, just like you would in a math course. Look over any sorts of charts, tables, and graphs too, and make sure you understand what they are meant to show.

Next, it's also important to take frequent breaks when studying, and to get some sleep. Research has shown that people generally have short attention spans and that studying for several smaller units of time is better than studying for fewer but larger units of time. In general, you shouldn't study for more than about 30 minutes without taking a significant break. Of course, using cards makes this easy, as you can grab a few and learn them, then do something else for a while, then grab a few other cards, etc.

Research also shows that much of our memory work is done in our sleep, and that we have a greater potential to remember things when we're well-rested. Studies have shown that studying for a few hours, with breaks, and then getting a good night's sleep is much more effective than "cramming" all night and getting little sleep.

Taking exams/quizzes:
First of all, get some sleep before an exam, and eat something before you come in! You should also plan on coming to class a little earlier on exam days so that you don't feel rushed to get in your seat, settle down, and start answering questions. Being late for an exam is also terribly distracting to other students that have already started. And, there is a strong correlation between students coming in late and forgetting to turn their phone off, which is not acceptable. So, get there early and try to just relax for a few minutes, or get in a few more minutes with your study cards.

All of the exams for this course have a time limit. Consistently a few students complain about the amount of time given, but those that study and know the material find that the amount of time given is significantly more than is needed. Basically, students that run out of time are typically the same students that would perform poorly on the same exam even if they had all the time they wanted. In other words, if you don't know the material – you don't know the material, and extra time won't make a difference. Besides, the CLEP, ACT, SAT, CPT, CLAS, FTCE, GRE, LSAT, MCAT, etc. exams are all timed, too. Even the ASVAB exam to get into the military is timed. In all of these cases, being able to answer questions in a timely manner is part of the test.

If you have trouble finishing exams in the time given, there are things you can do to help, though. For example, as you go through an exam, if you get to a question and can't remember the answer in a few seconds – move on. Do not sit and stare at it. That burns up time, and tends to erode your confidence for the rest of the exam. Instead, you should continue to move through the whole exam and answer every question you can answer in a timely manner. Then go back to anything you didn't answer.

Have you ever tried to remember the name of a person, or a song, etc. that you know that you know, but just can't seem to recall at that instant – then it popped into your head seemingly from nowhere minutes later? It happens to everyone at times, and tells us that some part of our brain was still working on finding that information even after the rest of our brain had moved onto something else. This is why you should move on if you can't answer a question quickly.

If you move on, the answer to a question you didn't answer may literally just pop into your head later. Hopefully before you finish your exam, of course. If that happens, go back and answer the question immediately. Likewise, a word, phrase, or picture, etc. that's part of another question sometimes triggers a memory. So, reading some other questions and answers can oftentimes help you recall the answer to one you didn't respond to.

And last, but not least, unless you're absolutely positive that you've made a mistake or careless error, do not change your answers. Studies have shown that over 80% of the time that students change an answer on an exam, they either change it from a right answer to a wrong one, or from a wrong one to another wrong one. So, when in doubt, it's better to just go with whatever you thought first.

That's it!

Unit 1

Section 1.1: The Universe, Our Solar System, and the Earth

A great amount of scientific evidence indicates that the Universe came into existence approximately **14 billion years ago**, when the "Big Bang" occurred.

It now contains at least a few hundred billion **galaxies**, which are large systems typically comprised of tens of billions to a few hundred billion stars and solar systems.

Galaxies also contain numerous **nebulae**, which are enormous clouds composed of various gases and tiny specks of dust produced by numerous stars that have already lived and died. These materials, produced by stars through various nuclear reactions, are scattered into space and form such clouds when large stars are destroyed in explosions as they die, which are called **supernovas**. New stars and solar systems can still form within these clouds though, as they contain enough material to produce several more generations of them.

Our own solar system, which formed approximately **4.5 billion years ago**, resides within the **Milky Way Galaxy**, and it includes the Sun and eight planets. These can be divided into the inner four terrestrial planets and the outer four Jovian planets, which are also known as the gas giants, as they are much larger than the terrestrial planets and have far larger atmospheres.

The **four terrestrial planets** are similar in structure, are composed primarily of rock and metal, and have relatively small atmospheres. These are: **Mercury, Venus, Earth, and Mars.**

The **four Jovian planets** are also similar in structure, but are composed primarily of gas and liquid with relatively small rock/metal cores. These are: **Jupiter, Saturn, Uranus, and Neptune**.

Our solar system also includes several **dwarf planets**, such as **Pluto**, and many moons. And, it includes many asteroids, meteoroids, and comets, as well as scattered dust and gas.

When it comes to the Earth, it can be divided into four "spheres", which are the:

Geosphere: Earth's rock/metal surface and interior.

Hydrosphere: Earth's water at and below the surface.

Atmosphere: Earth's enveloping layer of gas.

Biosphere: Earth's regions inhabited by living things.

Only on the first three of these spheres will be focused on in this course, though.

A typical spiral galaxy is composed of over one hundred billion stars.

A nebula is a huge cloud in space, composed primarily of dust and various gases.

The terrestrial planets: Mercury, Venus, Earth, and Mars.

The Jovian/gas giant planets: Jupiter, Saturn, Uranus, and Neptune.

Above, the terrestrial planets are shown to scale relative to each other, as are the Jovian plants. However, the two groups of planets are not shown to scale relative to each other. For example, Earth is the largest terrestrial planet with a diameter of 7,918mi, but Jupiter, the largest Jovian planet, has a diameter of 86,881mi!

Likewise, Pluto was called a planet, but its diameter is only 1,474mi - about half the distance across the U.S. So, it's now called a dwarf planet, along with several other similar bodies that orbit the Sun.

The dwarf planet Pluto.

Geosphere: Earth's rock-metal surface and interior.

Hydrosphere: Earth's water at and below the surface.

Atmosphere: Earth's enveloping layer of gas.

Section 1.2: The Earth's Atmosphere, Weather, and Climate

The Earth's Atmosphere:

The planets' atmospheres have varying compositions, and the Earth's **atmospheric composition is approximately:**

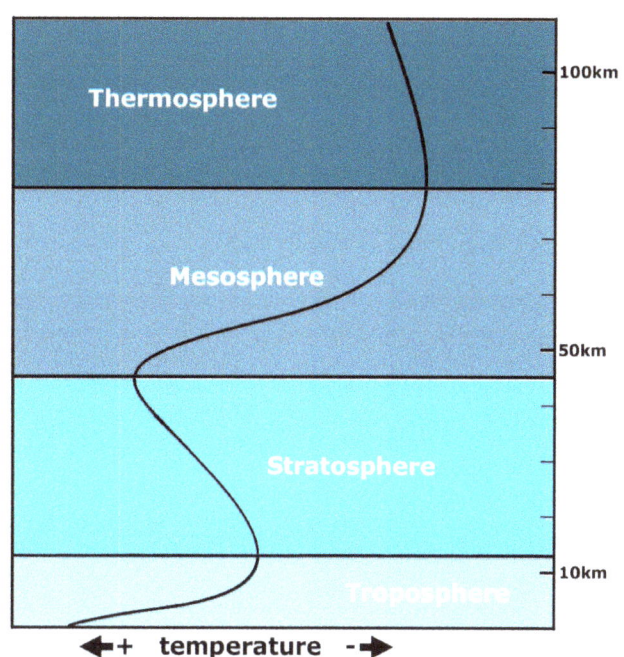

Temperature decreases or increases with altitude in the different layers of the atmosphere. Note that air temperature decreases as altitude increases in the Troposphere.

These percentages are relatively constant regardless of location, meaning they are consistent from place to place and at varying altitudes. However, the amount of water vapor (gaseous water) in the atmosphere is highly variable at different locations. This varying amount of water vapor present is what we call humidity.

The atmosphere also has several layers with distinct characteristics, which are the Troposphere, Stratosphere, Mesosphere, and the Thermosphere. The **Troposphere** is the lowest layer, and essentially all weather phenomena occur within it. Thus, it's the layer we are most concerned with.

The atmosphere is thicker around the equator and thinner near the poles, but **the Troposphere's average thickness is approximately:**

Atmospheric **air pressure** is the force exerted by the weight of the atmosphere at a given location. It varies significantly over time, and from location to location, but **the average air pressure at sea level is:**

Atmospheric pressure also **decreases rapidly** with increasing altitude, as the mass of air above a given point decreases. Air gets "thinner" as altitude increases, as well.

As altitude increases within the Troposphere, the air temperature generally decreases, too. The average change in temperature over a given distance is called the **normal lapse rate**, and *in the Troposphere, the normal lapse rate is:*

This means that the air temperature falls an average of **6.5ºC for each kilometer that altitude increases**, although it may change at a higher or lower rate under differing conditions. See the next page for an example.

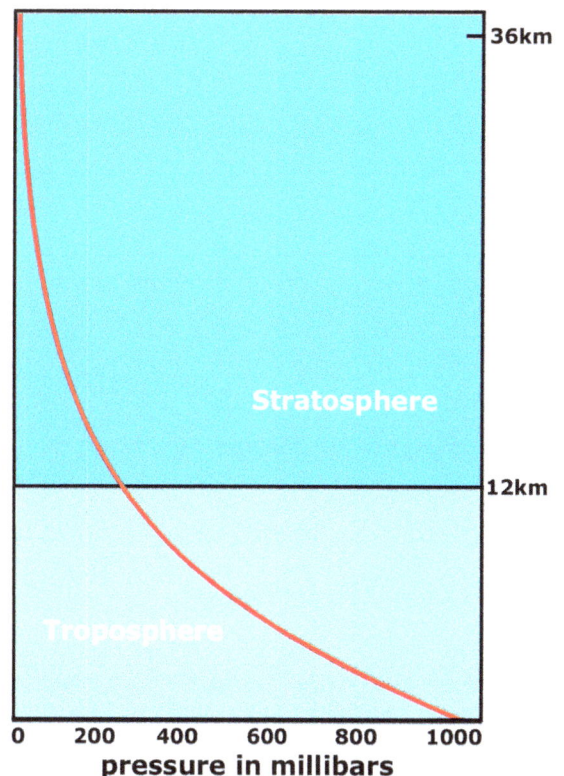

Air pressure also decreases as altitude increases, as there is a smaller and smaller mass of air above a given point.

An example of air temperature decreasing at the normal lapse rate:

For each increase of one kilometer in altitude, there is a corresponding 6.5ºC decrease in air temperature (on average).

Therefore, if the temperature at the base of a mountain is a cool 13ºC (or 55ºF), it would be approximately 6.5ºC lower at an altitude that's one kilometer higher than the base.

Thus, it would be a chilly 6.5ºC (or 44ºF).

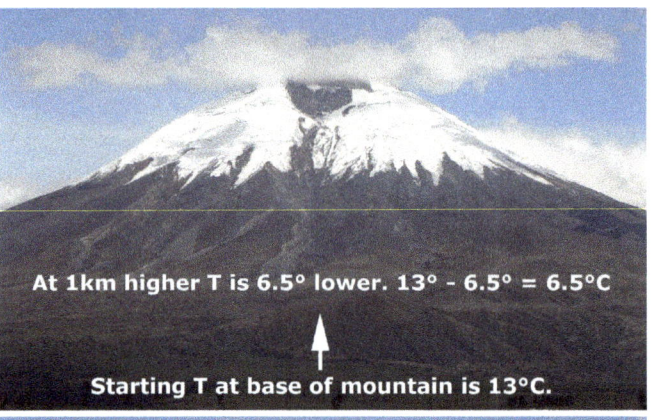

Likewise, by going up another kilometer, to an altitude that's two kilometers above the base of the mountain, the air temperature would be another 6.5ºC lower.

Thus, it would be a cold 0ºC (32ºF), which is the freezing point of water. Much of the mountain would likely be covered by snow/ice from this altitude up.

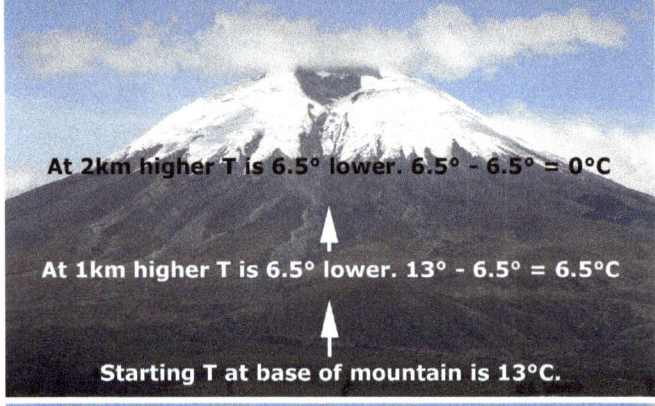

Continuing up another kilometer, to an altitude that's three kilometers above the base of the mountain, the air temperature would again be another 6.5ºC lower.

Thus, at the top of the mountain it would be an even colder -6.5ºC (20ºF).

This is why it's common to see snow and ice on the top of a tall mountain even if its base is snow-free.

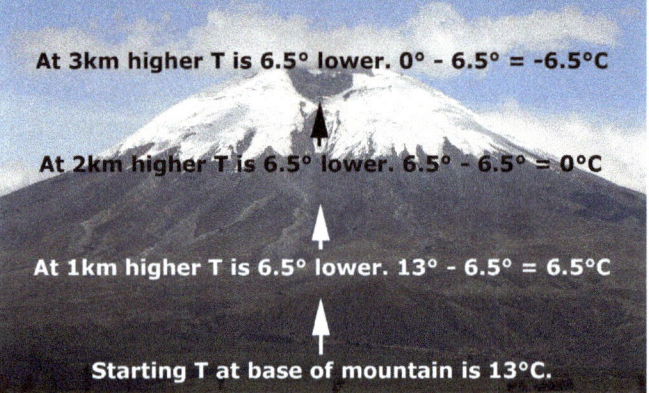

The Earth's Weather and Climate:

Weather is generally defined as:

It is considered to be **"short-term"**, as it can change significantly over relatively short periods of time.

Climate is generally defined as:

It is often thought of as **"long-term weather"**, as it refers to overall weather conditions over long spans of time, and does not change significantly over relatively short periods of time. Climate is sometimes said to be the sum of weather information that helps describe a place or region, as well.

The 6 basic elements of both weather and climate are:

Changes in all of these elements are driven by the **Sun's effects** on the Earth's surface and atmosphere. As will be covered, sunlight affects temperature, temperature affects humidity, and humidity affects cloudiness and precipitation. Temperature also affects air pressure, and air pressure affects winds.

Of course, when the Sun is up temperatures tend to rise, and when it is down, temperatures tend to fall. So, locations tend to heat up during the daytime and cool at night. **Several factors play a role** in this ever-changing swing of temperature, though.

#1 Temperatures are affected by the length of a given day/how long the Sun is up in any given 24 hour period. The longer the Sun is up, the more the Earth's surface and atmosphere heat up. So, longer days tend to be warmer, while shorter days tend to be cooler.

With the exception of the Equator, where the Sun is up for 12 hours and down for 12 hours every day of the year, the length of days at all locations on the Earth vary throughout the year. This is due to the **tilt of the Earth's axis** and the Earth's **position in its orbit** around the Sun.

All of the planets' axes are tilted, and the *Earth's axis is tilted:*

Thus, each of the Earth's hemispheres is tilted toward or away from the Sun at different positions in its orbit. This changes the hours of sunlight received, and is responsible for **seasonal changes** in weather.

When the Northern Hemisphere is tilted towards the Sun, the length of days is longer, temperatures rise, and it is summer. Conversely, when the Northern Hemisphere is tilted away from the Sun, the length of days is shorter, temperatures fall, and it is winter. Fall and spring are simply in between, when the Northern Hemisphere isn't tilted towards or away from the Sun. Also note that all of the seasons are opposite in the Southern Hemisphere, since it is tilted away from the Sun at the same time that the Northern Hemisphere is tilted towards the Sun, etc.

Example: When the Northern Hemisphere is tilted towards the Sun (as seen at position A on the diagram to the right), Tampa's days are approximately 14 hours long, temperatures rise, and it is summer. Six months later (position C), Tampa's days are about 10 hours long, temperatures fall, and it is winter. Tampa's days are 12 hours long at the points between these two extremes (positions B and D), temperatures are moderate, and it is either spring or fall.

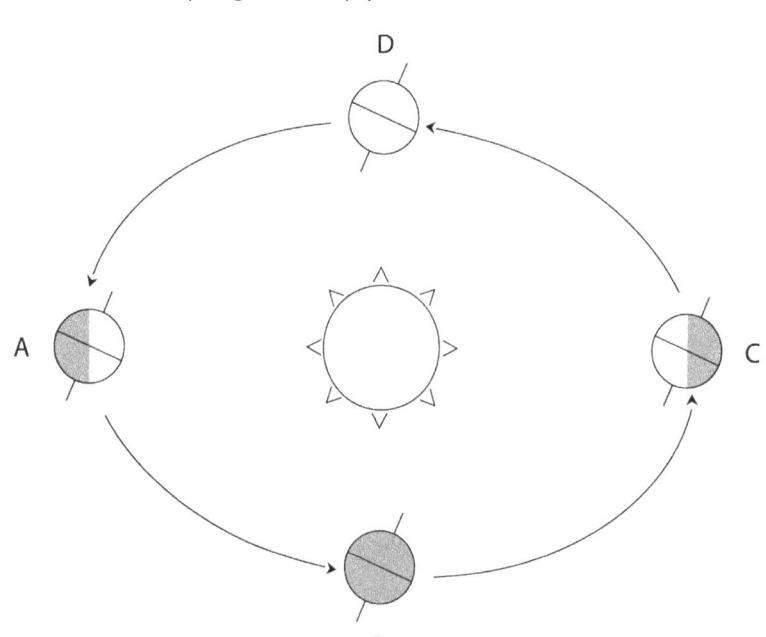

Position:	A	B	C	D
Name:				
Date of occurrence:				
Length of day:				

9

#2 Temperatures are also affected by the angle at which sunlight strikes the Earth's surface at different locations.

If the Sun is at a high angle in the sky (closer to straight up), then sunlight strikes surfaces at a high angle and can be thought of as being concentrated.

However, if the Sun is at a low angle in the sky (relatively far from straight up), then sunlight strikes surfaces at a low angle and can be thought of as being dispersed.

This is one reason that the Sun feels hotter in the middle of a given day, when it is highest in the sky, but cooler near sunrise and sunset, when it is lowest in the sky.

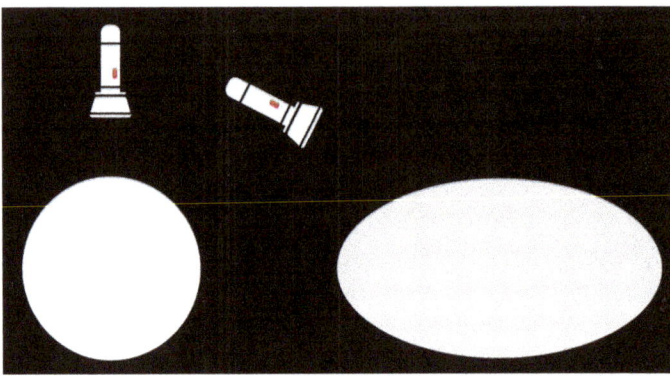

A good analogy is to think of a flashlight shining down onto a dark surface. If the flashlight is held directly over the surface and shining straight down, the light it emits will be concentrated and will warm the surface. However, if the flashlight is held at an angle, the light it emits will strike the surface at an angle and will be spread out more. Thus, a larger area would warm, but warm less.

High angle sunlight results in:

Low angle sunlight results in:

#3 Temperatures are also affected by the amount of air sunlight travels through.

Air is composed of gases, but also carries numerous aerosol particles (tiny airborne materials) such as dust, soot, ash, pollen, and bacteria. All of these can absorb, reflect, or scatter sunlight before it reaches the surface.

So, if the Sun is at a high angle in the sky and sunlight thus travels through less air before reaching the surface, more heating will occur.

Conversely, if the Sun is at a low angle in the sky and sunlight thus travels through more air before reaching the surface, less heating will occur.

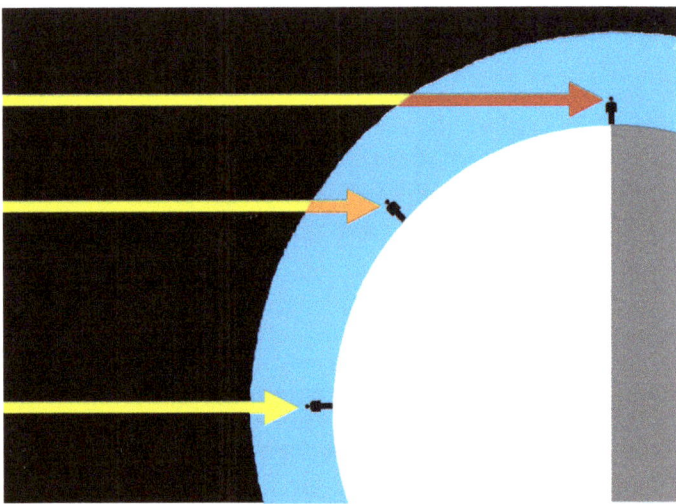

When sunlight is coming straight down through the atmosphere, it passes through less air than sunlight passing through the atmosphere at an angle. This, along with a low sun angle and the high albedo of snow and ice, is why the poles can remain cold all year.

Traveling through less air results in:

Traveling through more air results in:

#4 Temperatures are also affected by surfaces' albedo.

The term **albedo** refers to how reflective surfaces are, and **relatively dark surfaces have a low albedo**. So, things like rocky mountains and fields of dirt absorb much sunlight, and this heats them up.

Conversely, relatively **light surfaces have a high albedo**, like clouds, fields covered by snow, and areas covered by ice sheets, and reflect much sunlight. Thus, they resist heating up.

Sunlight striking a low-albedo surface results in:

Sunlight striking a high-albedo surface results in:

#5 Temperatures on land can also be affected by bodies of water.

Bodies of water resist temperature changes to a much greater extent than land areas do, and thus do not heat and cool as quickly. Therefore, while temperatures on land may vary greatly from summer to winter, the temperature of the oceans varies far less through the year.

Air often flows from over the oceans onto land masses. So, in the summer, relatively cool air may move from over an ocean onto land and cool the land. Conversely, in the winter relatively warm air may move from over an ocean onto land and warm the land. And, the closer a location on land is to an ocean, the greater the effect. The same occurs with other bodies of water, such as lakes, but to a lesser degree due to their smaller size.

For example, Seattle and Spokane are both in Washington, and are at the same latitude. However, Seattle is much closer to the west coast, and thus the Pacific Ocean, while Spokane is about 300 miles inland and on the opposite side of the Cascade Mountains.

In the summer, the Pacific Ocean stays relatively cool. So, when air moves from the ocean onto Washington in the summer, it cools Seattle. And, in the winter the Pacific Ocean stays relatively warm. So, when air moves from the ocean onto Washington in winter, it warms Seattle. This gives Seattle a mild climate for being so far north, with a reduced difference in its summer and winter temperatures.

On the other hand, air coming onto Washington from the Pacific Ocean must travel significantly farther to reach Spokane, and is affected by the temperature of the land it passes over on the way. Thus, Spokane isn't cooled as much in the summer, or warmed as much in the winter. It therefore has significantly warmer summers, and significantly cooler winters.

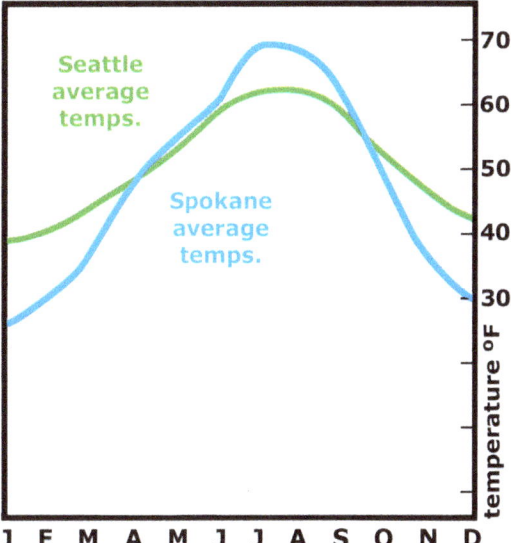

A simple climograph that shows average monthly temperatures for Seattle and Spokane. Notice that Seattle is warmer than Spokane in the winter months, and is also cooler than Spokane in the summer months.

Section 1.3: Humidity, Clouds, and Precipitation

When water changes state, from solid to liquid or liquid to gas, etc., the local environment around it is either warmed or cooled.

The melting of ice is a:

The freezing of water is a:

The evaporation of water (change from liquid to gas) is a:

The condensation of water (change from a gas to a liquid) is a:

When ice melts in water, the water is cooled.

Humidity:

Humidity is the amount of water vapor in a given amount of air.

And, any device used to measure humidity is called a:

The humidity of a given amount of air can be determined and presented in different ways. However, only two are significant here, which are specific humidity and relative humidity.

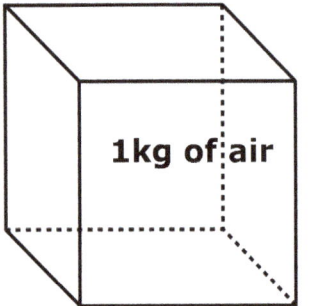

Specific Humidity:

Units used:

Example: If a box containing 1kg of air was found to be containing 20g of water vapor, the specific humidity would be:

Any given amount of air can only hold a certain amount of water vapor, with that amount being **dependent upon the air's temperature**. If a given amount of air is holding the maximum amount of water vapor that it can possibly hold at a given temperature, then it is **saturated**, or full. Conversely, if a given amount of air is holding less than it can possibly hold at a given temperature, it is **unsaturated**.

Water Vapor Capacity:

Air Temperature	W.V. Capacity
40°C	47g/kg
30°C	28g/kg
25°C	20g/kg
20°C	14g/kg
15°C	10g/kg
10°C	7g/kg
5°C	5g/kg
0°C	3.5g/kg
-10°C	2g/kg
-20°C	0.75g/kg
-30°C	0.3g/kg
-40°C	0.1g/kg

Units used:

Example: A box containing 1kg of air with a temperature of 20°C would have a water vapor capacity of:

If the temperature of the air rises, its water vapor capacity will also:

If the temperature of the air falls, its water vapor capacity will also:

Relative humidity:

Units used:

Example: A box containing 1kg of air with a temperature of 25°C would have a water vapor capacity of 20g/kg (from the chart on page 12). If the box contained 10g of water vapor, the air's specific humidity would be 10g/kg. Thus, its relative humidity would be:

1kg of air **0g water vapor** **@ 20°C** **20g liquid water**	**1kg of air** **7g water vapor** **@ 20°C** **13g liquid water**	**1kg of air** **14g water vapor** **@ 20°C** **6g liquid water**

Above is a hypothetical situation combining the concepts of specific humidity (SH), water vapor capacity (WVC), saturation, and relative humidity (RH) by using a box holding 1kg of air at 20°C and 20g of liquid water. Note that at 20°C, the water vapor capacity (WVC) is 14g/kg.

At first, the WVC is 14g/kg, but there is no water vapor in the air. Thus, the air has a SH of 0g/kg, a RH of 0%, and is unsaturated. Because the air is unsaturated, over time, some of the 20g of liquid water in the box will evaporate and change state to water vapor.

Later, some evaporation has taken place, and the air now holds 7g of water vapor. Thus, the air has a SH of 7g/kg, a RH of 50% (the amount of vapor present is 50% of the WVC), and is still unsaturated. Because the air is still unsaturated, over time, some of the remaining 13g of liquid water in the box will evaporate and change state to water vapor.

Even later, more evaporation has taken place, and the air now holds 14g of water vapor. Thus, the air has a SH of 14g/kg, a RH of 100% (the amount of vapor present is 100% of the WVC), and is now saturated. Because the air is now saturated, no more evaporation will take place. The SH will remain at 14g/kg, and the amount of liquid water will remain at 6g.

The only way for more water to evaporate at this point would be to warm the air and increase the WVC, causing the air to become unsaturated again.

Continuing with relative humidity, as stated, it is also affected by temperature. This is because relative humidity is affected by water vapor capacity, and water vapor capacity is affected by air temperature.

If air is saturated and has a relative humidity of 100%, but is then warmed, that would then increase the water vapor capacity and the air will become unsaturated. This means the relative humidity would fall to less than 100%.

Conversely, if air has a relative humidity of 50%, but is then cooled, that would then decrease the water vapor capacity and the air would then be closer to saturation. This means the relative humidity would rise to greater than 50%.

Likewise, if air has a relative humidity of 100%, but is then cooled, that would then decrease the water vapor capacity and the air would then be holding more water vapor than is possible. So, some of the water vapor would have to change state and condense into liquid water, leaving the air. The relative humidity would stay at 100%, but there would be less water vapor present.

Condensation occurs when water vapor changes state to liquid water. This happens when air's water vapor capacity falls below its specific humidity, which often occurs when warm humid air contacts a cold surface/object, like this soda can.

Above is another hypothetical situation combining the concepts of specific humidity (SH), water vapor capacity (WVC), saturation, and relative humidity (RH) by using a box holding 1kg of air at 20°C. However, in this situation, the temperature of the air will fall over time. Note that at 20°C, the water vapor capacity (WVC) is 14g/kg.

At first, the WVC is 14g/kg, and there is 7g of water vapor in the air. Thus, the air has a SH of 7g/kg, a RH of 50%, and is unsaturated.

Later, the temperature has fallen to from 20°C to 10°C. The air still holds 7g of water vapor, but due to the cooling, the WVC had fallen to 7g/kg. Thus, the air has a SH of 7g/kg, a RH of 100%, and is saturated.

Even later, the temperature has continued to fall from 10°C to 0°C. Because the air continued to cool after it was saturated, condensation occurred. The air was holding 7g of water vapor and had a SH of 7g/kg, but the WVC at 0°C is only 3.5g/kg. So, 3.5g of the water vapor had to leave the air via condensation, and change state to liquid water. This would leave the SH at 3.5g/kg, the RH at 100%, and the air saturated. However, there would now be 3.5g of liquid water in the box, as well.

Dew Point:

Units used:

Example: A box containing 1kg of air with a temperature of 20°C would have a water vapor capacity of 14g/kg. If the box contained 7g of water vapor, the air's specific humidity would be 7g/kg, and its relative humidity would be 50%. The air would become saturated and the relative humidity would rise to 100% if the temperature were to fall to:

Cloud Formation:

If a gas is allowed to expand, its temperature will fall. Likewise, if a gas is compressed, its temperature will rise. Therefore, if a parcel of air rises to a higher altitude in the atmosphere, where air pressure is lower, it will expand and cool. And, if a parcel of air falls in the atmosphere, to a lower altitude where pressure is higher, it will be compressed and will warm.

Adiabatic temperature change:

So, if a parcel of warm, humid air rises in the atmosphere, it will cool and its relative humidity will rise. If it rises enough, it will eventually become saturated when it reaches the "dew point in the sky". Then, if it rises any further, condensation must occur and some of the water vapor in the air will change state to liquid water. This is how almost all clouds are formed, as many clouds are composed of tremendous numbers of microscopic droplets of liquid water that have condensed from air. Likewise, clouds may also contain large numbers of tiny ice crystals if the air inside the cloud falls to a low enough temperature to create them, often through the process of **deposition**. When deposition occurs, water vapor changes state directly from a gas to a solid (ice), skipping the liquid phase that typically occurs in between.

Thus, the key to cloud formation, whether they're made of liquid water, ice crystals, or a mixture of both, is for humid air to cool past its dew point. Again, **this most often occurs when air rises**, and there are four basic mechanisms/processes that can cause air to rise.

#1 Frontal lifting/wedging:

Warm air is less dense than cold air. So, when parcels of warm air and cold air meet along warm and cold fronts, the warm air always rises over the cold air.

If the warm air is humid, clouds will begin to form at the **condensation level**, which is the altitude at which the dew point is reached.

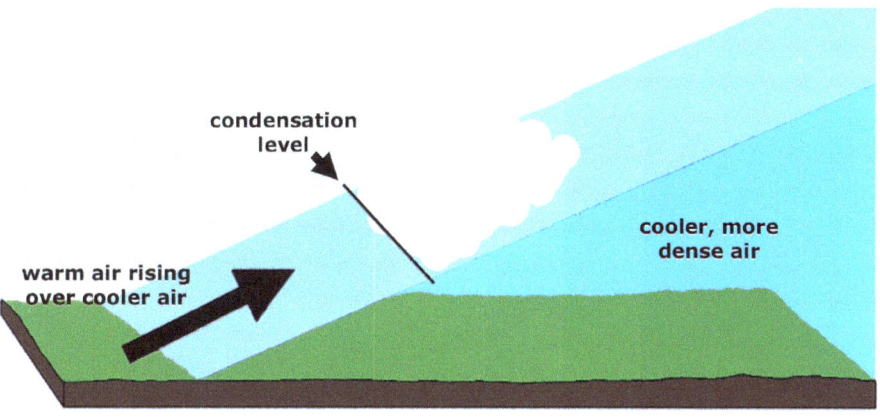

#2 Convergent lifting:

In Florida, especially in the summer, warm, humid air moves onto land from both the Gulf of Mexico and the Atlantic Ocean, and then meets over the state.

The air thus converges and rises, creating many clouds, and plenty of summertime rain.

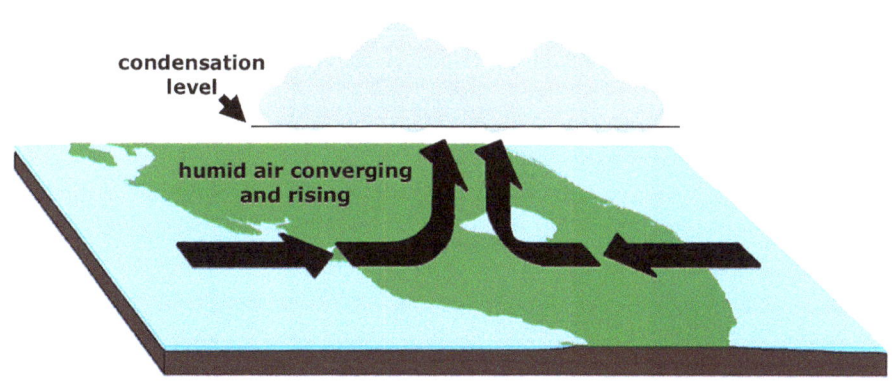

#3 Convective lifting:

In Florida, especially during the summertime, the land is warmed by sunlight, which then causes humid air over the land to warm and rise.

This typically creates lots of cotton-ball-like clouds, with flat-bases.

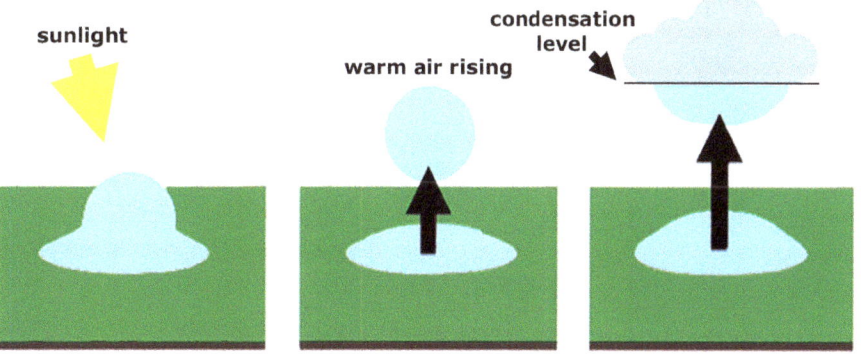

#4 Orographic lifting:

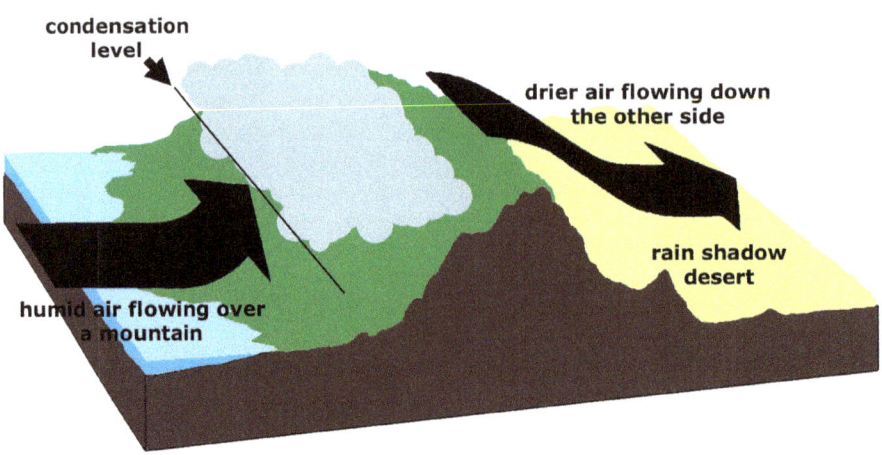

While mountains often cause clouds to form, it is common for clouds to form and produce rain or snow only on one side, leaving the air drier as it passes over the top.

Then, as the now dry air descends on the other side of a mountain, a rain shadow desert may be created.

Basic Cloud Types:

There are numerous types of clouds, with about a dozen being common. In general, they are classified by their overall form and the altitude at which they are formed/found. The basic forms are cirrus, cumulus, and stratus, with the altitudes being high, mid, and low, with exceptions.

Form #1 Cirrus:

Form #2 Cumulus:

Form #3 Stratus:

Cirrus (or **Cirro**) is also used for high-altitude clouds.

High-altitude #1 Cirrus:

High-altitude #2 Cirrocumulus:

High-altitude #3 Cirrostratus:

Alto is used for mid-altitude clouds.

Mid-altitude #1 Altocumulus:

Mid-altitude #2 Altostratus:

Stratus (or **Strato**) is also used for low-altitude clouds.

Low-altitude #1 Stratus:

Low-altitude #2 Stratocumulus:

Low-altitude #3 Nimbostratus:

Low-altitude #4 Fog:

Some clouds do not fit into the basic altitude ranges, as they can grow taller with development and may simultaneously be low, mid, and high-altitude. These are **clouds with vertical development**, called **cumulonimbus clouds**, that often reach from close to the ground up to the top of the Troposphere.

The change in atmospheric conditions at the boundary of the Troposphere typically causes such clouds to stop growing at this altitude and flatten out on top. Strong high-altitude winds often affect their form as well, and lead to the formation of **anvil-top cumulonimbus clouds**.

The rapid growth of a cloud with vertical development, over a period of just 20 minutes.

CwVD Cumulonimbus:

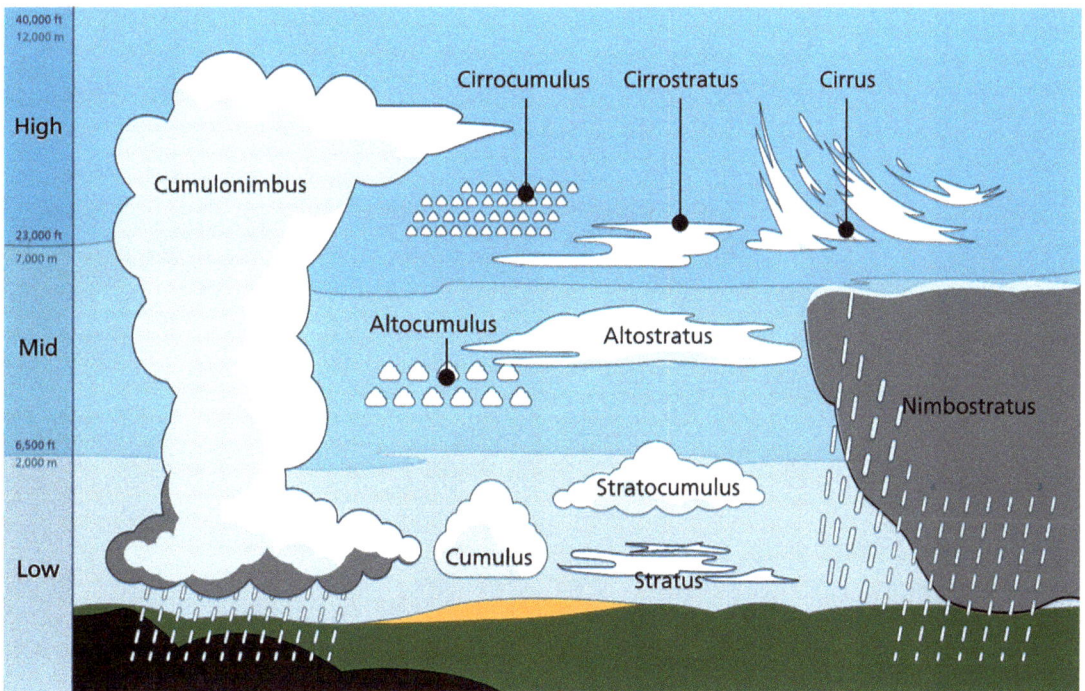

All of the basic cloud types.

Precipitation:

Precipitation of various sorts is produced when water vapor condenses in the atmosphere and falls to the ground as liquid water or ice. It comes is several common forms, each of which is produced under differing atmospheric conditions.

Drizzle and Rain:

Snow:

Sleet:

Graupel:

Hail:

CwVD Cumulonimbus:

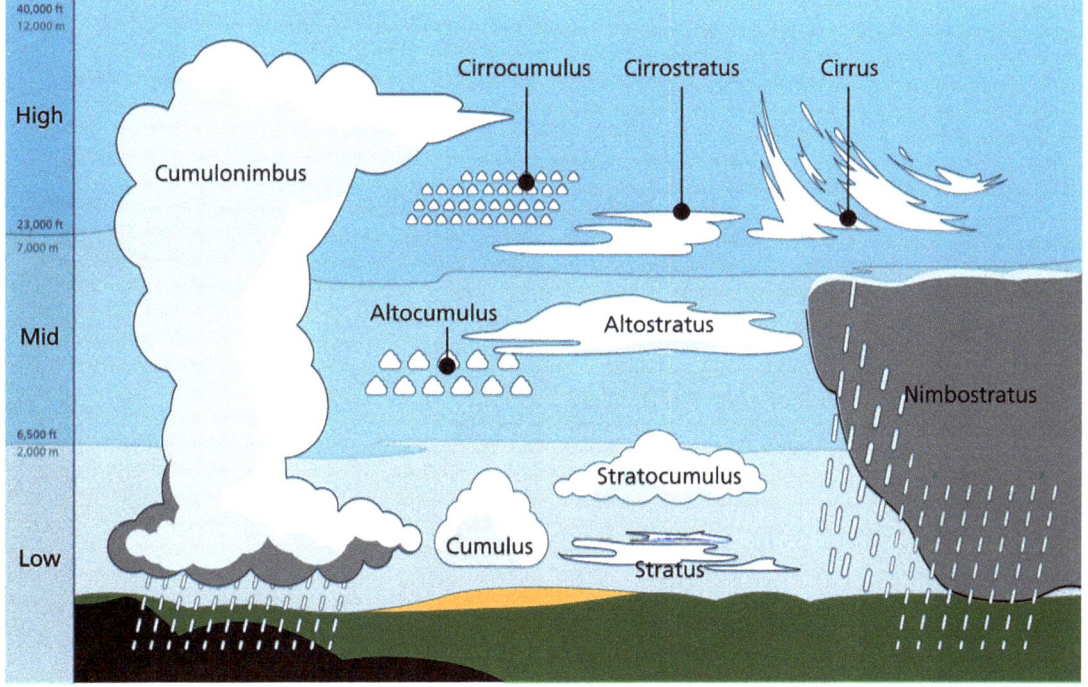

All of the basic cloud types.

Precipitation:

Precipitation of various sorts is produced when water vapor condenses in the atmosphere and falls to the ground as liquid water or ice. It comes is several common forms, each of which is produced under differing atmospheric conditions.

Drizzle and Rain:

Snow:

Sleet:

Graupel:

Hail:

While technically not forms of precipitation, there are three other things to note. When condensation takes place upon surfaces at ground-level, **dew** is formed. Likewise, under very cold ground conditions, ice may form on surfaces at ground-level through deposition and create **frost** on them. Also, when rain freezes upon contact with very cold surfaces at ground-level, a coating of clear ice can be produced, which is known as **glaze**. When glaze forms, the rain itself it typically called freezing rain.

Dew develops on surfaces when humid air at ground-level falls below its dew point.

Frost develops when water vapor is deposited onto cold surfaces.

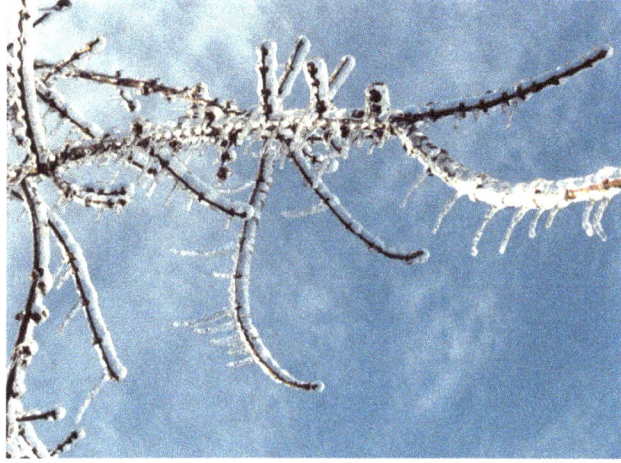

Glaze forms when rain freezes upon coming into contact with very cold surfaces.

Section 1.4: Air Pressure and Wind

Again, atmospheric **air pressure** is the force exerted by the weight of the atmosphere at a given location. It is typically measured as millibars (metric) or inches of mercury (U.S. Customary), and varies significantly over time, and from location to location, with ***the average air pressure at sea level being:***

And, any device used to measure air pressure is called a:

Air Pressure and Wind:

Heating air causes it to expand and become less dense. Thus, heating a parcel of air will result in a decrease in air pressure. Likewise, cooling air causes it to contract and become more dense. So, cooling a parcel of air will result in an increase in air pressure.

Because Earth's surface is heated unevenly, with some areas being warmed while others cool, some areas will be covered by relatively low pressure air with others being covered by relatively high pressure air. These **differences in pressure from location to location create wind**, as air always flows from areas of higher pressure to areas of lower pressure. However, there are three things that primarily affect the nature of winds.

#1 The pressure gradient:

Pressure gradients create wind, and the greater the gradient, the stronger the wind.

Pressures may be shown on weather maps as numbers, but are often shown on maps as isobars, as well. **Isobars** are lines that connect areas of equal pressure on a map, and the relative spacing between these lines can be used to determine the pressure gradient, and thus approximate wind velocities.

Closely-spaced isobars:

Well-spaced isobars:

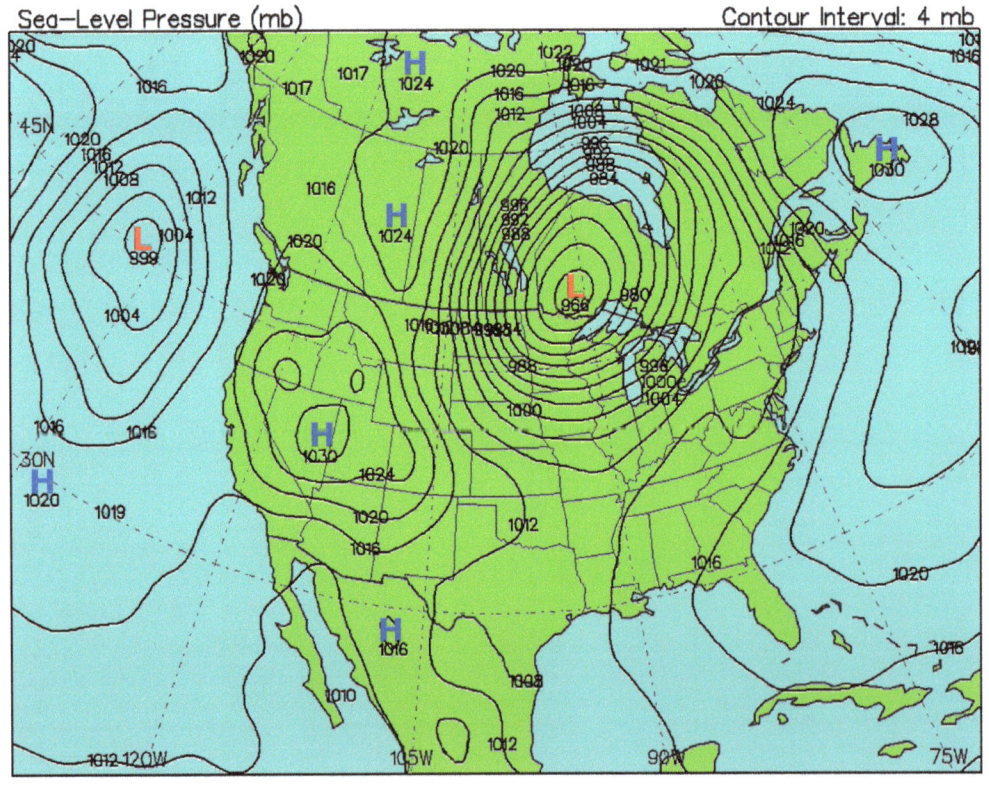

This weather map shows areas of high and low air pressure (blue Hs and red Ls), pressure in millibars at specific locations, and isobars with an interval of 4mb.

Winds will be faster in an area where the isobars are relatively closely-spaced, indicating a high pressure gradient, such as around the area of low pressure seen in the middle portion of the map. And, winds will be slower where the isobars are relatively far apart, indicating a low pressure gradient, as seen at the lower portion of the map.

As air attempts to flow perpendicular to isobars, directly from areas of relatively high pressure into areas of lower pressure, the Earth's spin affects its direction of travel. This turning, or deflection, of air is the second factor that affects the nature of winds.

#2 The Coriolis Effect:

Winds in the Northern Hemisphere are deflected:

Winds in the Southern Hemisphere are deflected:

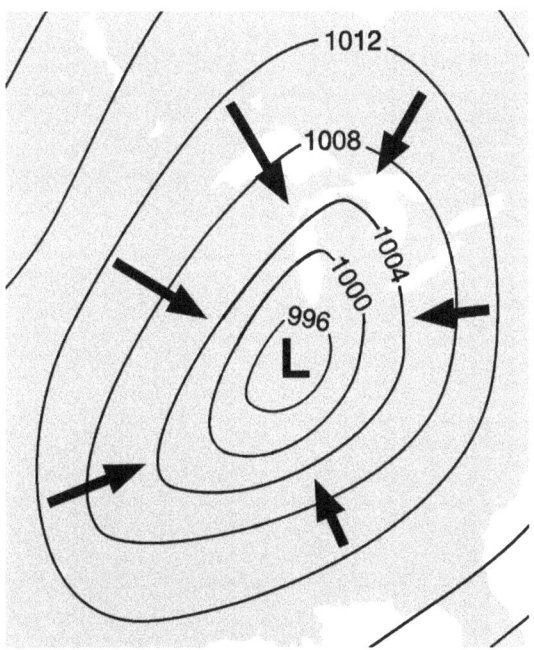

If the Earth didn't spin, air would flow directly from an area of higher pressure into an area of lower pressure.

Due to the Coriolis Effect, tropical cyclones in the Northern Hemisphere rotate in a counter-clockwise direction as air spirals into them.

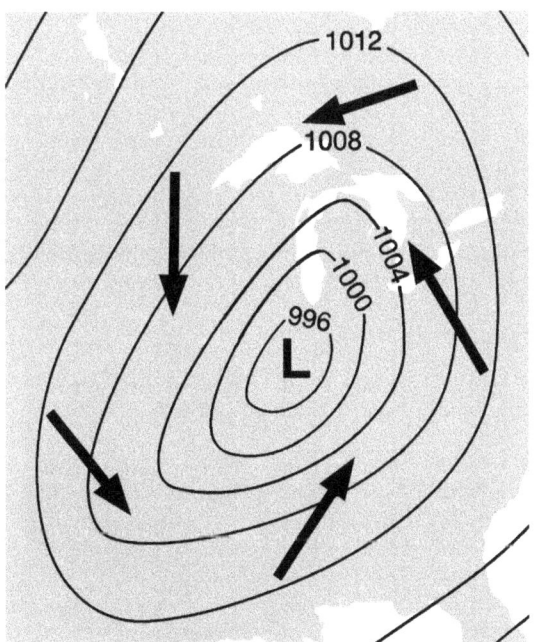

The Earth does spin though, so the Coriolis Effect causes a deflection of winds, which then spiral inwards.

Conversely, those in the Southern Hemisphere rotate in a clockwise direction.

The Coriolis Effect also:

Lastly, as air flows from one place to another, it may be in contact only with other air, or with water (such as the surface of an ocean), or with land. All of these create friction or drag, and can dramatically affect wind velocity. Thus, the third factor that affects that nature of winds is friction.

#3 Friction:

High and Low Pressure Systems:

When areas of low or high pressure develop, air moves into those with low pressure and out of those with high pressure. This can lead to the formation of **cyclones** and **anticyclones**, respectively.

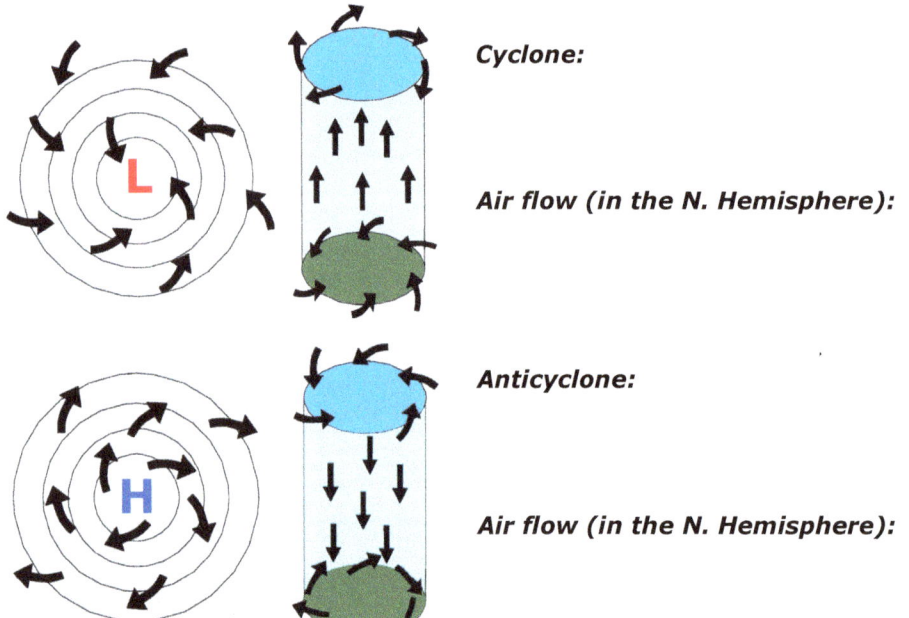

Cyclone:

Air flow (in the N. Hemisphere):

Anticyclone:

Air flow (in the N. Hemisphere):

Convergence occurs in cyclones, with surface air flowing in and up, and cooling. So, cyclones typically produce clouds and precipitation, as the rising air cools. Conversely, divergence occurs in anticyclones, with relatively cold, dry air moving down, warming, and flowing out. This warming increases the water vapor capacity of already dry air. So, anticyclones do not create clouds, but instead result in clear skies.

Both are also affected by the flow of air at high altitudes. Fast-moving currents of air flow through the Troposphere at altitude, which are called **jet streams**, and air flow within them can speed up or slow down.

When air flow in a jet stream speeds up, divergence occurs as the air quickly moves out of an area. This reduces air pressure and draws more air in from below. This helps to maintain a low-pressure system as the air flows in and up at the surface.

On the other hand, when airflow in a jet stream slows down, convergence occurs as the air "piles up", and it then moves downward towards the surface. This helps to maintain a high-pressure system as the air flows down and out at the surface.

The Polar (blue) and Subtropical (red) jet streams.

General Air Circulation:

Air circulation on a global scale is driven primarily by **convection**, as air is warmed in some areas, becomes less dense and rises, then cools at altitude, and becomes more dense and sinks.

Near 0°, warm humid air converges and rises, creating the Equatorial Low (also known as the Intertropical Convergence Zone), which is a band of low pressure around the equator. This air then cools and creates clouds and precipitation, leaving the air drier.

The air then diverges aloft and moves poleward, until it sinks and creates the **Subtropical Highs around 30°N and 30°S**, which are bands of high pressure around these latitudes. The air warms as it descends, but it is dry, and clouds generally are not produced.

Air also converges, rises, and creates the **Subpolar Lows around 60°N and 60°S**, which are bands of low pressure around these latitudes. Some air moves polewards aloft and then sinks, producing the **Polar Highs at around 90°N and 90°S**, and some air moves towards the equator and sinks at the Subtropical Highs.

As descending air flows out from the Subtropical Highs, it diverges and attempts to flow north and south, but is deflected by the Coriolis Effect. This creates the Trade Winds and Westerlies. The **Trade Winds exist from near the equator to around 30° north and south, and blow from east to west**, while **the Westerlies exist from around 30° to 60° north and south** and **blow from west to east**.

As air sinks and diverges at the poles, it is also deflected by the Coriolis Effect. This produces the **Polar Easterlies**, which exist from **around 60° to 90° north and south** and **blow from east to west**.

So, air moves primarily from east to west at low latitudes, from west to east and mid-latitudes, and east to west at high latitudes. The U.S. is right in the middle, and this is one of the reasons we experience such variable weather.

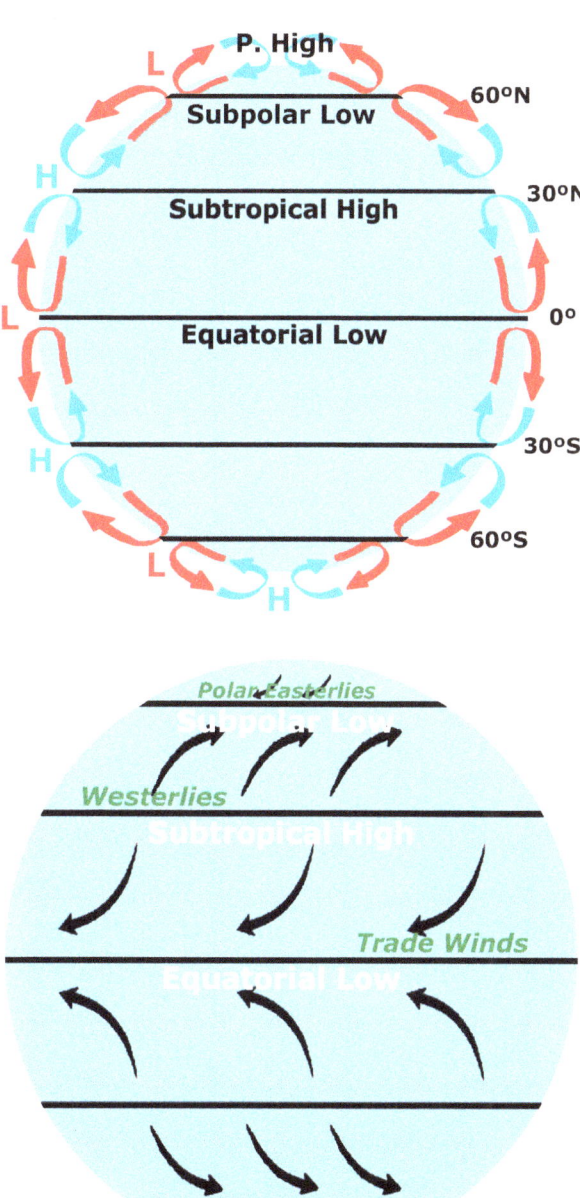

The Equatorial Low/ITCZ is visible, where large numbers of clouds and heavy precipitation are produced daily.

Section 1.5: Air Masses and Weather Fronts

Air Masses:

An **air mass** is a large parcel of air with horizontally uniform temperature, humidity, and pressure. They develop in different areas, move across the Earth's surface, and also interact with each other.

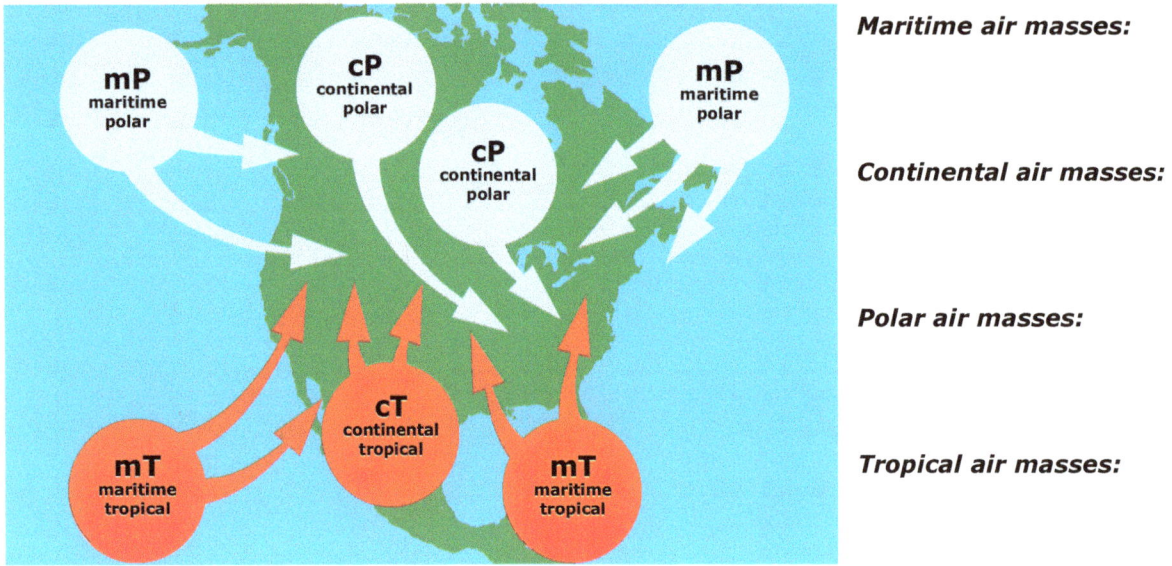

Maritime air masses:

Continental air masses:

Polar air masses:

Tropical air masses:

In combination, an air mass may be maritime polar, continental polar, maritime tropical, or continental tropical, each of which has distinct characteristics.

mP:

cP:

mT:

cT:

Weather Fronts:

As air masses move across the Earth's surface, a weather front is created when they interact. So, **fronts** are areas where two or more air masses meet, and there are four types.

#1 Warm Front:

Symbol:

Pressure change:

Weather produced:

#2 Cold Front:

Symbol:

Pressure change:

Weather produced:

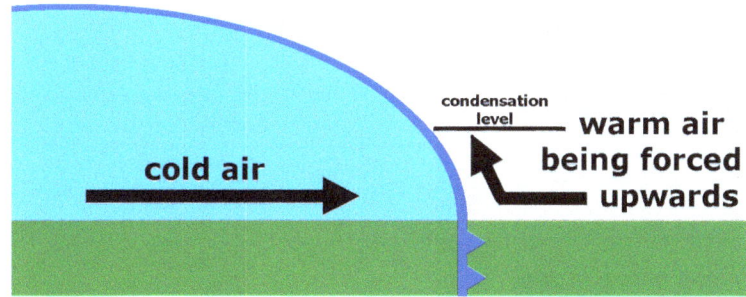

#3 Stationary Front:

Symbol:

Pressure change:

Weather produced:

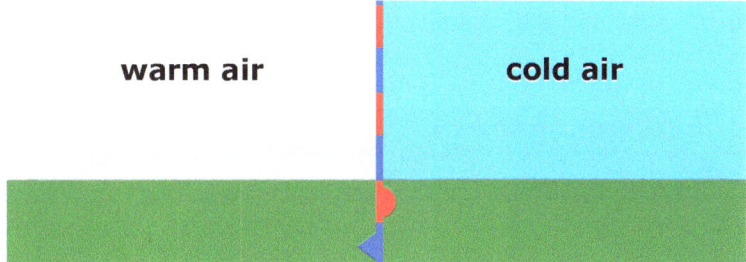

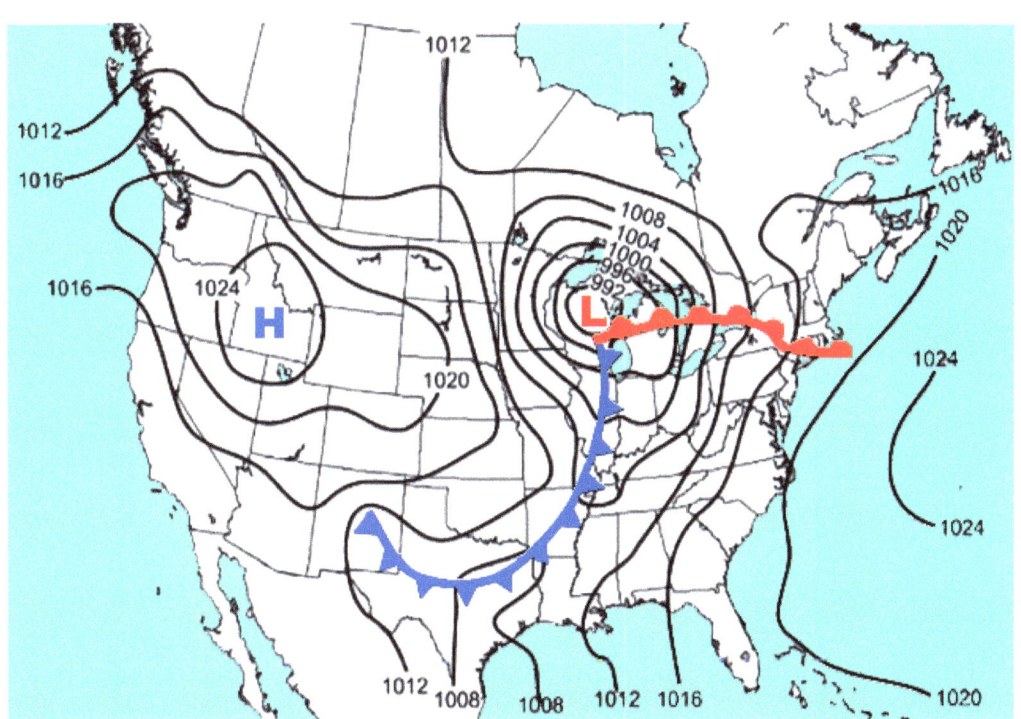

This weather map shows areas of high and low air pressure, pressure in millibars at specific locations, and isobars with an interval of 4mb. It also shows a cold front being produced by a cP air mass moving down from Canada, and a warm front being produced by a mT air mass moving up from the Atlantic/Gulf of Mexico.

#4 Occluded Front:

Symbol:

Pressure change:

Weather produced:

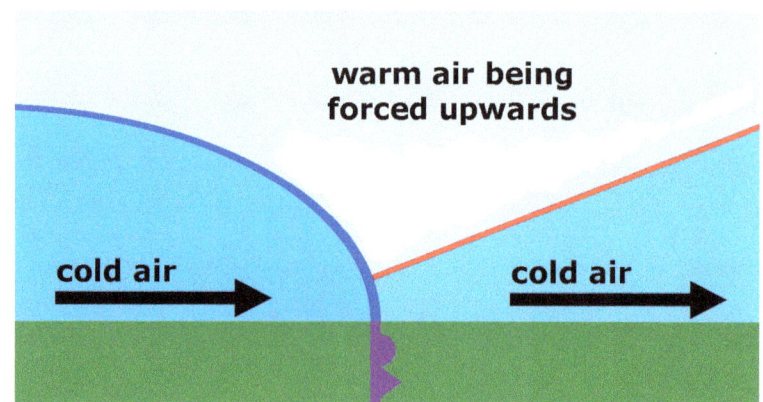

Occluded fronts are typically produced within **mid-latitude cyclones** (also called extratropical cyclones), which are large scale cyclonic weather systems that form at middle latitudes. Many pass over the U.S. each year, and are responsible for much of the severe weather we experience, especially in the winter. In fact, the "tail" that typically hangs down from these **number-nine-shaped systems**, which is produced by a cold front, sometimes reaches all the way to Florida, delivering heavy rains in the winter.

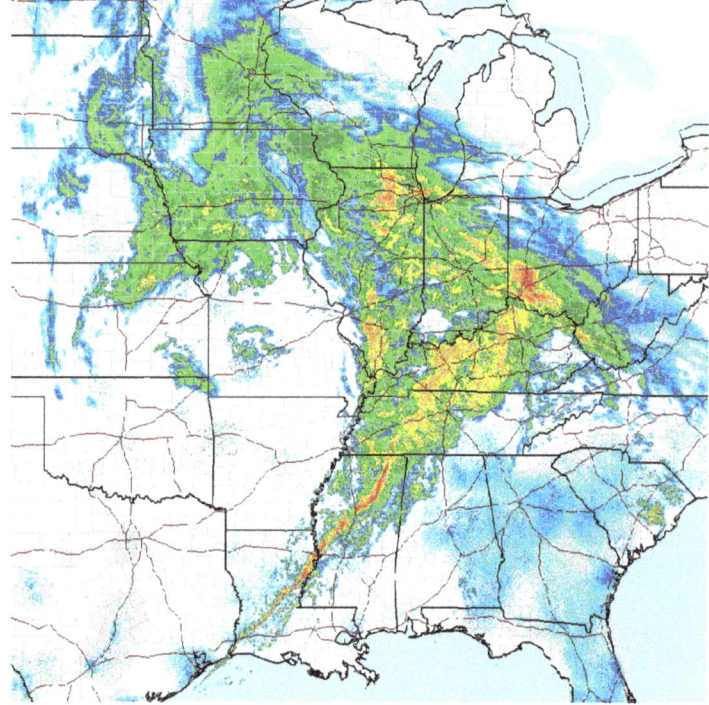

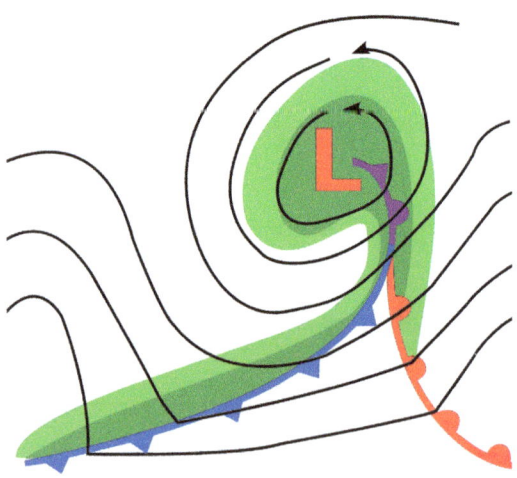

Above are two images of mid-latitude cyclones, which take on a number-nine form in the Northern Hemisphere. The above-left image is from a satellite and shows the clouds formed by one. And, the above-right image is a weather radar composite that shows the precipitation created by another. These systems can clearly produce severe weather over large areas.

Mid-latitude cyclones develop when cold dry air moves southward and begins to rotate along with warm humid air moving northward. So, initially a cold and warm front form, which are connected to each other. As rotation continues, faster moving cold air can twist underneath the warm front and create an occluded front within the core of the system, as seen in the image to the left.

Section 1.6: Severe Weather

The term **severe weather** refers to any potentially dangerous meteorological phenomena that can cause property damage, serious social disruption, and/or threaten human health. Ice storms, blizzards, dust storms, and such can all be considered severe weather events, but the best known and most commonly thought of are thunderstorms, tornadoes, and hurricanes.

Thunderstorms:

A **thunderstorm** is any storm that produces lightning and thunder. They are **produced by cumulonimbus clouds**, which typically produce lightning/thunder, strong winds, heavy rain, and sometimes hail. They may also produce tornadoes. Being clouds with vertical development, these start as low-level cumulus clouds that grow to immense sizes under particular atmospheric conditions.

As a reminder, condensation is a warming process. So, when unsaturated air rises, it cools at a given rate, but will cool slower if it becomes saturated and condensation occurs. This can create **unstable air**, which is air that is warmer than the environment around it, and can quickly produce large clouds.

Basically, when a rising parcel of air is trying to cool from rising and expanding, it can also be trying to warm at the same time if condensation occurs within it. Therefore, if there is sufficient humidity in the air, and significant condensation occurs within it as it rises, it will cool at a lower rate than the environment around it. This means the parcel of air will be warmer and less dense than the surrounding air, and will continue to rise, as a hot air balloon would. If this occurs, more air can be drawn upwards from below, which can lead to more condensation, more warming, and the continued growth of a cloud. Such conditions are called **atmospheric instability**, again when a parcel of air rises, but remains warmer than the environment around it.

Regardless, the life cycle of a thunderstorm/cumulonimbus cloud has **three stages**.

#1 Cumulus Stage:

#2 Mature Stage:

#3 Dissipating Stage:

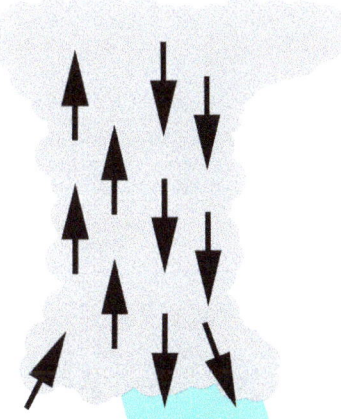
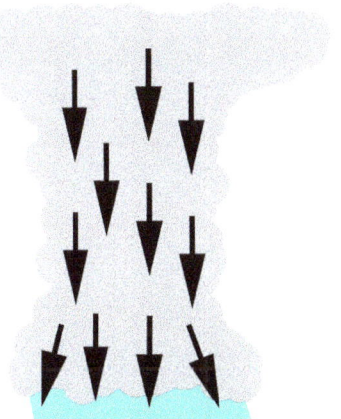

During the Cumulus Stage, updrafts draw more humid air into the cloud. This leads to more condensation, and the cloud rapidly grows upwards.

During the Mature Stage updrafts continue in part of the cloud, while precipitation creates downdrafts in other parts.

During the Dissipating Stage updrafts stop, but the precipitation continues to create downdrafts. So, the cloud eventually "rains out".

Thunderstorm hazards:

Key West, flooded by a large thunderstorm.

A car damaged by large hail.

Tornadoes:

A **tornado** is a violent windstorm that extends from a cumulonimbus cloud to the surface in the form of a vortex. They are usually preceded by a **funnel cloud**, which then contacts the ground, and may last from a few minutes to a few hours. They have the potential to cause significant harm, as well.

Pressure inside the vortex is very low, so air rapidly spirals into a tornado at ground level and moves up into the cloud overhead, typically carrying suspended debris. Such debris may be slung away aloft and scattered widely, too.

Also note that a **tornado watch** is issued if an area is under conditions appropriate for tornado formation, but a **tornado warning** means a funnel cloud or tornado has been spotted or detected by radar.

Typical size range of tornadoes:

Average size of a tornado:

Average ground speed of a tornado:

Average wind speed in a tornado:

Maximum wind speed in a tornado:

Tornadoes may form at any time of the year, but **the majority occur in the spring and early summer** months. During this period, maritime-tropical air masses move over the U.S. from the Gulf of Mexico and collide with continental-polar air masses moving down from Canada. These air masses, at this time, tend to have the greatest difference in temperature and humidity, which leads to the formation of especially strong thunderstorms. So, this is when conditions are best for tornado formation.

When tornadoes do form, they're ranked on the **Enhanced Fujita Scale**, based on their estimated wind speed.

Approximately **65% of all tornadoes are ranked as EF0 to EF1**, and do little, if any serious damage. However, approximately **33% are ranked at EF2 to EF3**, and can do significant damage. **Only 2% of all tornadoes are ranked at EF4 or EF5**, yet these are responsible for approximately 65% of all deaths. So, they are uncommon, but still cause the most harm.

Tornado hazards:

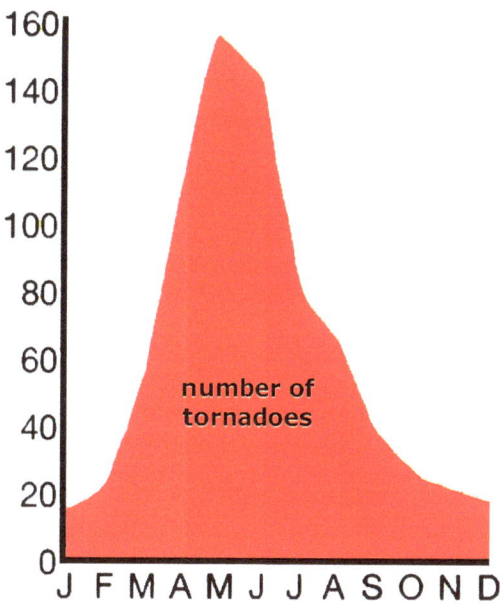

Enhanced Fujita Scale	
EF Rating	3 second gust (mph)
0	65-85
1	86-110
2	111-135
3	136-165
4	166-200
5	over 200

Tropical Cyclones:

Tropical cyclones are rapidly rotating storm systems characterized by a low-pressure center, a spiral arrangement of thunderstorms, heavy rains, and strong winds. Depending on their location of occurrence, they are commonly called **hurricanes** (the Atlantic, Gulf of Mexico, and Northeast Pacific), **typhoons** (Northwest Pacific), or just **cyclones** (Southern Hemisphere).

Unlike tornadoes, fully developed tropical cyclones may be hundreds of miles across and can last for several days (31 days is the record). So, they can cause damage over very large areas. While the most powerful tornadoes may have wind speeds over 300mph, large tropical cyclones may have wind speeds over 200mph, which can still be devastating.

There are **three stages** in the development of a these systems, which are dependent on atmospheric pressure and the resultant wind speeds.

#1 Initially tropical cyclonic weather systems begin as **tropical depressions** with maximum sustained wind speeds at the surface being **less than 39mph**.

#2 If intensification occurs, these become **tropical storms** with wind speeds from **39 to 73mph**.

#3 With further intensification, these become **tropical cyclones** with wind speeds of **74mph or higher**.

If an area of low pressure forms over **warm water** in the tropics and air pressure in the area is low enough, a tropical depression can develop. Air will flow inwards and spiral up, as it does in any cyclonic weather system. This leads to the formation of many cumulonimbus clouds, as the air moving over warm surface water will have a high humidity.

As more air flows inwards, causing increased condensation and cloud formation, local air pressure can continue to fall. This will create a chain-reaction, as an even lower air pressure will cause warm, humid air to spiral into the area faster and faster. Thus, a tropical depression can become a tropical storm if sustained winds speeds reach **39mph** or higher.

If there is sufficient humidity and air pressure continues to fall, cloud formation will continue to increase, as will wind speeds. If sustained winds reach **74mph** a tropical storm has become a tropical cyclone, and the process can continue. Wind speeds in the strongest of these may eventually reach over 200mph.

While air converges at the surface and spirals upward, it also diverges aloft, flowing away from the central

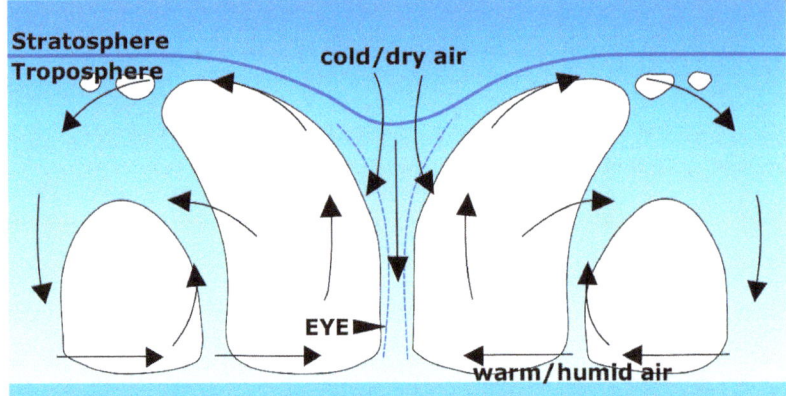

Strong outflow at the top of a tropical cyclone is responsible for the formation of an eye, which may be up to 50 miles across and nearly cloud free.

The well-developed eye of a large typhoon.

area of the system. As this outflow aloft increases, cold dry air is drawn downwards into the system's center, which prevents cloud formation. This leads to the development of an **eye**, an area at the center of a tropical cyclone that winds rotate around, but which itself is relatively calm.

Also, if a tropical storm becomes a tropical cyclone, in the U.S. they are then ranked on the **Saffir-Simpson Hurricane Wind Scale**. This scale is based on wind speed, which increases as air pressure falls in the middle of the system. The lower the pressure, the higher the wind speed, and the higher ranking they are given on the scale.

Storm Classification	Wind Speed (mph)
Tropical Depression	0-38
Tropical Storm	39-73
Hurricanes	
Category 1	74-95
Category 2	96-110
Category 3	111-130
Category 4	131-155
Category 5	>155

The Saffir-Simpson Hurricane Wind Scale. Note on the right that wind speed increases as air pressure falls.

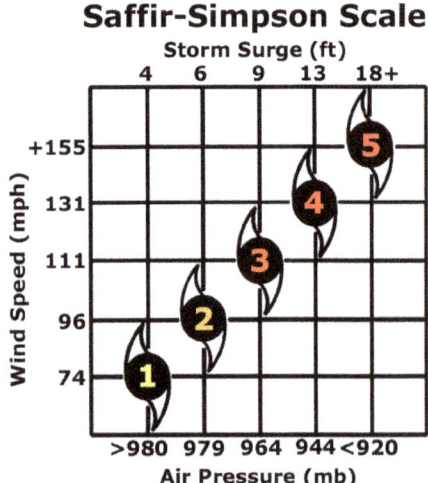

For tropical cyclones to develop, **water temperatures need to be above 80°F**, which is why there is a **hurricane season** in the U.S. It runs from **June through November**, as water temperatures are typically warm enough for tropical cyclones to form during this time. They have formed outside of this hurricane season, but it is very uncommon.

Also, **with the exception of the South Atlantic and the Southeast Pacific**, tropical cyclones typically form in only two zones, **between 5° and 20° north and south**, and then move westward in the Trade Winds.

Tropical cyclones typically develop only within these zones because:

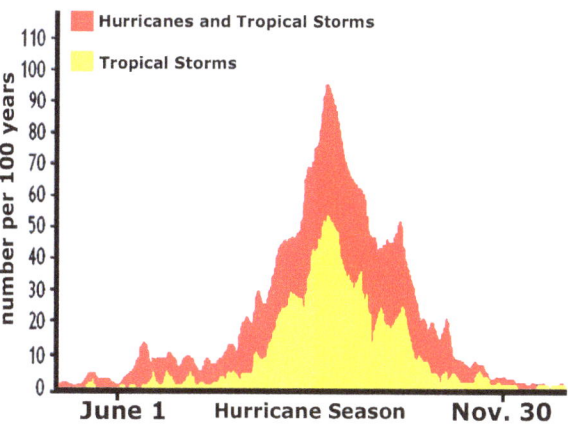

Once developed, it is common for tropical cyclones to leave these zones and continue to much higher latitudes. They also tend to curve towards higher latitudes, and may dramatically change their direction of travel **if they move into the Westerlies**. This is why they may hit Florida from both the east and the west. Eventually they dissipate though, which again, may take many days or even weeks.

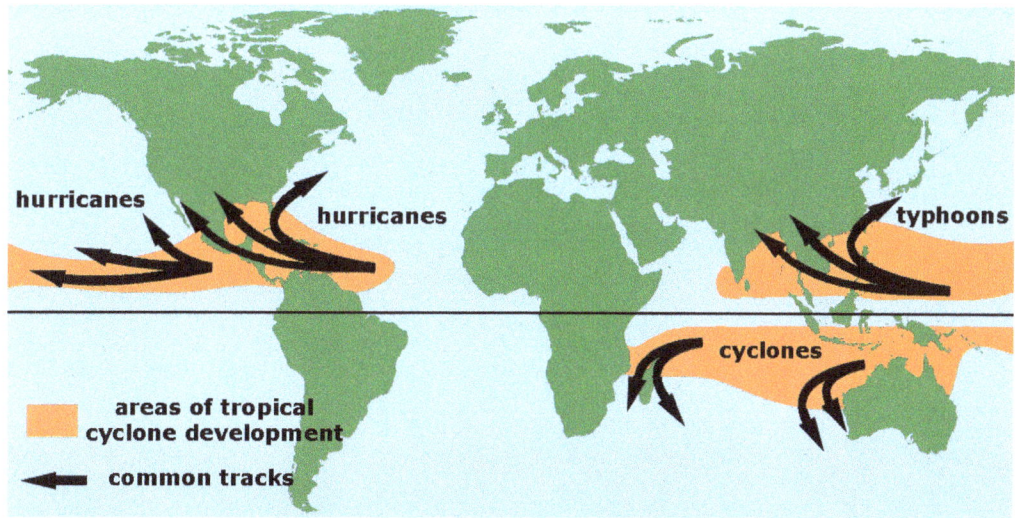

The areas where tropical cyclone formation occurs globally.

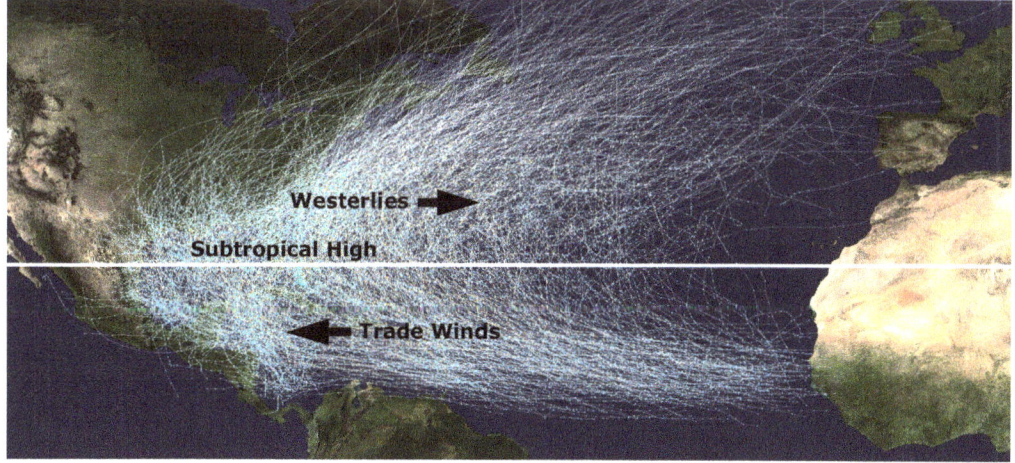

Notice that many hurricanes re-curve eastward as they pass into the Westerlies.

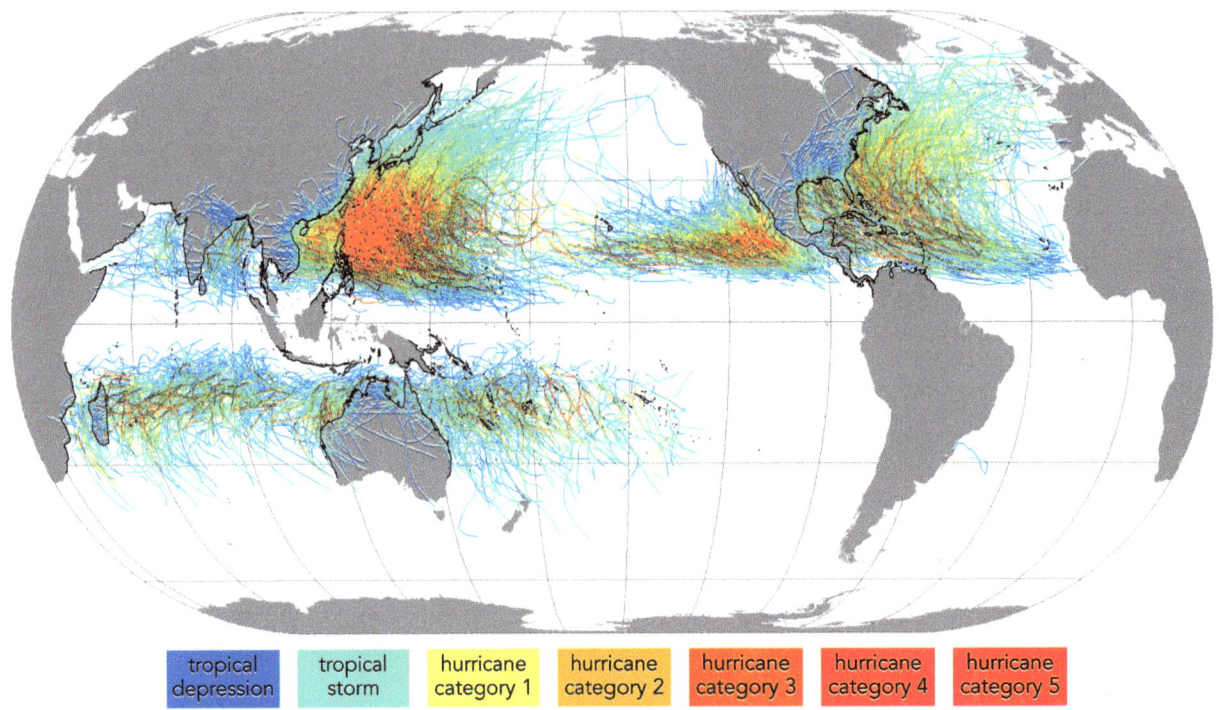

Sixty years of tropical cyclonic weather activity. Taking everything covered into consideration, it is easy to see why they form and travel where they do.

Tropical cyclones dissipate when:

Tropical cyclone hazards:

Unit 2

Section 2.1: Minerals and Rocks

A **mineral** is a naturally occurring, solid, inorganic substance, and a **rock** is a naturally occurring solid mass made of one or more minerals. There are about 4,000 known minerals and several hundred different rocks, which is too many for any one person to know. However, there are relatively few minerals and rocks that are abundant. So, only the defining characteristics of minerals and how to identify them, along with the most common rocks/types of rocks, will be covered.

Minerals:

All minerals have the following five defining characteristics:

When it comes to identifying minerals, each has some particular qualities/properties that make it distinct. There are many ways mineral specimens can be examined or tested, but the following are the basic characteristics used to identify them, all of which are expanded upon in the lab manual.

Minerals have a **luster**, which is the appearance of light reflected from its surface. There are numerous specific types of luster, but the broadest division is between those that have a metallic luster and those that have a non-metallic luster. Minerals that have a metallic luster look like they are composed of metal, and those with a non-metallic luster don't appear to be composed of metal.

Pyrite is a mineral with a metallic luster.

Hornblende has a non-metallic luster.

Minerals come in different **colors**, which may be useful in identification. However, the color of one mineral can be variable at times, and many different minerals can have the same color. Therefore, their color oftentimes is not very useful.

Minerals have a **hardness**, which is a measure of how well a specimen can resist being scratched by another material. This depends on how strongly the atoms in a mineral are bonded together, and some minerals are very hard (like diamond), while others are quite soft (like talc), with many more that are in between (like fluorite). So, minerals are ranked on a ten-point scale called **Mohs Hardness Scale**. To test a mineral's hardness, a specimen is either used to try to scratch something with a known hardness, such as a piece of glass, or something with a known hardness is used to try to scratch the specimen.

Mineral	Scale Number
Talc	1
Gypsum	2
Calcite	3
Fluorite	4
Apatite	5
Orthoclase	6
Quartz	7
Topaz	8
Corundum	9
Diamond	10

Minerals can develop a distinctive **crystal form** under the right conditions, such as cubic, rhombohedral, or hexagonal. However, under other conditions crystals either can't form or are deformed, etc. And, a few minerals can come in more than one crystal form, while many different minerals can produce similar crystals. So, at times crystal form may useful in indentifying minerals, but at other times it may not.

Some common crystal forms are cubic (L), rhombohedral (M), and hexagonal (R). Pictured are fluorite, calcite, and quartz.

Some minerals have **cleavage**, which means they can break along planes of weakness in their crystalline structure. The chemical bonds that hold the atoms in a crystal together may be weaker in some places than others, and if planes of weakness exist, a mineral is said to have cleavage. This means that it can be cleaved (split) in some fashion. However, if a mineral does not have cleavage it will instead break into randomly-shaped pieces that oftentimes have sharp or jagged edges. This is called **fracture**.

Mica has good cleavage and can be split into thin sheets.

Sphalerite is dark, but makes a light streak.

Some minerals have a distinctive **streak**, which is their color in powdered form. If a small amount of a mineral specimen is ground into powder, it would seem like the powder should be the same color as the whole specimen. However, in some cases the powder can be very different in appearance. To test a mineral's streak, a small ceramic tile can be used, which is called a streak plate. When a specimen is rubbed across a plate some of it grinds off and this will typically create a colored streak.

Some minerals react with **acid**, and will effervesce (fizz) if acid is dropped on them. Some minerals can be dissolved by sulfuric acid, but others may be unaffected, while some may be dissolved by hydrochloric acid, etc. The most common use of an acid test is to verify whether a specimen is made of calcium carbonate ($CaCO_3$), as it effervesces when tested with dilute hydrochloric acid.

Some iron-containing minerals are **magnetic** and will pick up a paper clip like a magnet would, while others will only pick up or stick to a magnet. A magnetic mineral will also attract the needle of a compass, which is a common test.

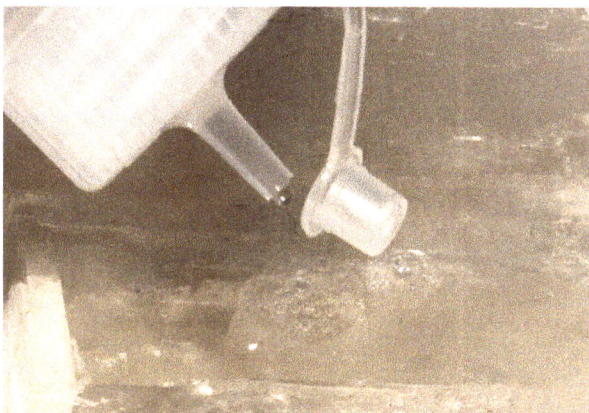

Calcite ($CaCO_3$), and any rock made of it, will effervesce with the addition of hydrochloric acid.

Lastly, there are many other characteristics that can be useful at times, such as taste, feel, smell, elasticity, malleability, fluorescence, and radioactivity.

Rocks:

Again, a **rock** is a naturally occurring solid mass made of one or more minerals. There are three basic rock types, and hundreds of named rocks, but almost all of them are made of differing combinations of the same eight elements. Of these eight, two dominate the rest, which are **oxygen** and **silicon**. So, most minerals and rocks contain both oxygen and silicon, with **oxygen being the single most abundant element in them**. Any mineral or rock that contains both is called a **silicate mineral** or a **silicate rock**, and silicates make up **about 90% of the Earth's crust**. Conversely, those that do not contain both oxygen and silicon are called **non-silicates**.

The three basic rock types are:

Igneous rocks:

Igneous rocks are split into either **intrusive rocks**, which form below ground and cool slowly, or **extrusive rocks**, which form at the surface and cool quickly. They're also classified/identified according to their **texture** and **composition**. The texture of an igneous rock is simply its overall appearance, specifically the average crystal size it displays and/or the presence of vesicles (bubbles).

The five igneous textures are:

If molten rock is allowed to cool underground and solidify very slowly, then large crystals typically form, which are easily visible to the naked eye. Conversely, if molten rock cools relatively quickly, coming out of a volcano, the crystals will generally be either too small to see with the naked eye, or no crystals will form at all (exceptionally rapid cooling). So, most igneous rocks either have large crystals and a **coarse-grained texture**, small crystals and a **fine-grained texture**, or no crystals and a **glassy texture**.

However, there are times when molten rock may solidify at two distinct rates. Imagine magma cooling very slowly underground leading to the formation of some large crystals, but then being ejected during volcanic activity and finishing the solidification process very quickly, leading to the formation of small crystals. Such a rock would thus have some large crystals that are mixed in a background, or matrix, of small crystals. Any such igneous rock, having two distinct crystal sizes, has a **porphyritic texture**.

Lastly, molten rock may contain large quantities of dissolved gas, which will attempt to escape if brought near or to the surface. If the magma/lava has a sufficiently low viscosity and cools slowly enough, these gasses can escape to the atmosphere. However, at other times the gasses cannot escape before solidification occurs, and the rock produced will contain significant numbers of trapped bubbles. Any such igneous rock having a significant number of bubbles in it has a **vesicular texture**.

The composition of molten rock is variable for a number of reasons, and igneous rocks may in turn be very light to very dark in overall color. **Granitic igneous rocks** have a relatively high percentage of light-colored silicate minerals and low percentage of dark silicate minerals, so they are relatively light in color. Significant quantities of dark crystals may be present, but a granitic rock's overall appearance is still light. Conversely, **basaltic igneous rocks** contain a relatively high percentage of dark silicate minerals and low percentage of light silicate minerals, so a basaltic rock's overall appearance is dark. And, **andesitic igneous rocks** have an intermediate composition between these two, and are intermediate in overall appearance.

Igneous Rocks			
Texture	light	intermediate	dark
coarse-grained	granite	diorite	gabbro
fine-grained	rhyolite	andesite	basalt
porphyritic	porphyritic rhyolite	porphyritic andesite	porphyritic basalt
vesicular	pumice		scoria
glassy		obsidian	

The two most common granitic rocks are:

    intrusive   extrusive

The two most common basaltic rocks are:

    intrusive   extrusive

The two most common andesitic rocks are:

    intrusive   extrusive

The two most common vesicular rocks are:

    light   dark

Sedimentary rocks:

Sedimentary rocks are classified/identified according to **how they form**, **their grain size**, and **their composition**. So, they're initially split into two categories based on how they form, which are detrital sedimentary rocks and chemical sedimentary rocks.

Detrital sedimentary rocks are composed primarily of sediments derived from other rocks, meaning they're made of rock particles derived from pre-existing rocks that have been broken down into pieces. Such rocks are classified/identified by the average size of the sediments they're comprised of, which is called their **grain size**.

The four grain sizes of detrital sedimentary rocks are:

The process of turning sediments into sedimentary rocks is called **lithification**, and this is how detrital sedimentary rocks are formed. Sediments can be buried, compacted by the weight of additional overlying sediments, and then lithified via **cementation**, which is the gluing together of grains. Cementation can occur as mineral-laden water soaks through the sediments, and the cements that glue them together are typically either **calcite** or **quartz**.

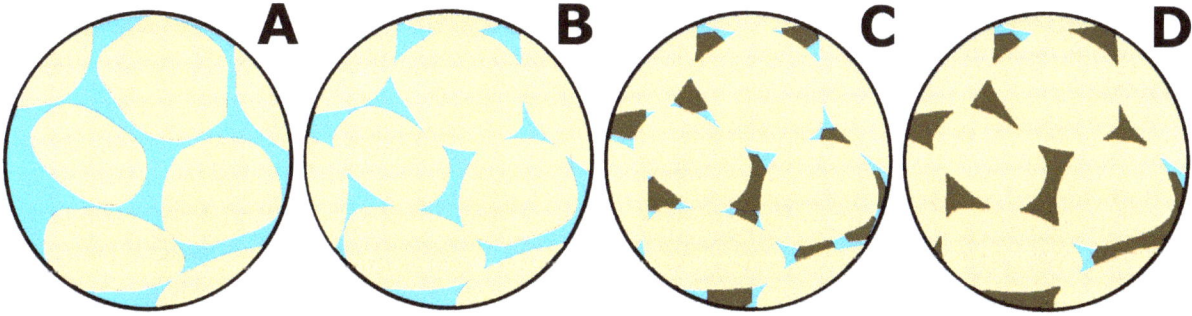

The stages of lithification. A) Sediments are deposited and water soaks through the pore spaces between them. B) The sediments are compacted. C) Typically, calcite or quartz cement begins to form in the pore spaces between them. D) Cement continues to form in between them, and may eventually fill all pore spaces.

On the other hand, **chemical sedimentary rocks** are formed when substances are dissolved away from pre-existing rocks into water and then precipitated from that water to form new rock. This process can occur in a number of different ways, making classification/identification less straightforward. In general they're classified by composition though, both chemical composition and the type of particles of which they're comprised.

Many chemical sedimentary rocks are composed of calcium carbonate that was produced by various organisms, including shell material, coral skeletons, and calcitic plankton skeletons. However, some are made of siliceous (quartz) plankton skeletons, or plant material. Other chemical sedimentary rocks can form abiotically though, meaning their constituents are not produced by living things. These simply precipitate from water under certain conditions, especially when mineral-laden water evaporates and leaves mineral deposits behind. Such rocks are called **evaporites**, as they form due to evaporation.

Detrital Sedimentary Rocks	
Description	Rock Name
coarse grain size = gravel (angular grains)	breccia
coarse grain size = gravel (rounded grains)	conglomerate
medium grain size = sand	sandstone
fine grain size = silt	siltstone
very fine grain size = mud	shale

Angular grains have:

Rounded grains have:

Calcitic rocks:

Siliceous rocks:

Chalk is:

Chert is:

Flint is:

Chemical Sedimentary Rocks	
Description	Rock Name
abiotic calcite	limestone
cemented shells, corals, fragments, etc.	fossiliferous limestone
calcitic plankton shells	chalk
loosely cemented calcitic shells and fragments	coquina
siliceous plankton shells and abiotic silica	chert / flint
halite formed when water evaporates	rock salt
gypsum formed when water evaporates	rock gypsum
compressed and lithified plant matter	coal

41

Metamorphic rocks:

The metamorphism of rocks occurs primarily in **two settings**. Some metamorphic rocks are formed in a **baked zone** where magma bodies heat surrounding rocks underground, but most are created by the immense heat and pressure generated during **mountain building**, when mountains are produced by plate tectonic activity.

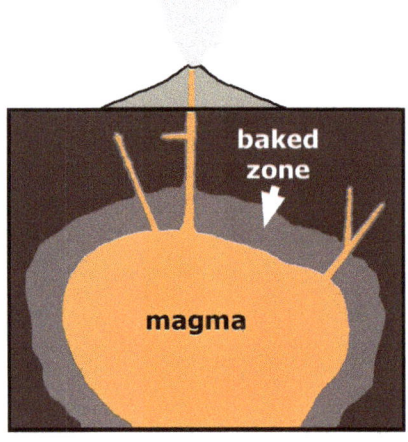

When metamorphism occurs, the degree of alteration of a rock is variable, depending on how much heat/pressure a rock is subjected to. So, metamorphism can be low-grade to high-grade. When a rock undergoes low-grade metamorphism it will typically look much like the **parent rock** (original rock). However, high-grade metamorphism can alter the appearance of a rock to such a degree that it may look significantly different and bear little resemblance to the parent rock. Others may be in between, as well.

Metamorphic rocks are classified/identified according to **their texture**, **grain size**, and **composition**. There are only two metamorphic textures though, which are **foliated** and **non-foliated**. Foliated metamorphic rocks are produced if any crystals or particles in a rock are re-aligned under high temperature/pressure conditions, which leads to the development of a layered or banded appearance. Conversely, non-foliated metamorphic rocks are produced if such crystals/particles do not form layers or bands. This typically occurs when a parent rock is composed of a single mineral.

Some rocks are altered by bodies of magma as they melt their way towards Earth's surface.

Grain size, as was the case with the other rock types, is simply the average crystal size of a specimen. Metamorphic rocks can thus be fine-grained, medium-grained, or coarse-grained depending on how large its crystals/particles are.

Lastly, non-foliated metamorphic rocks, which are typically composed of a single mineral, can be further classified by their **mineral composition**. These are typically made of **calcite** or **quartz**.

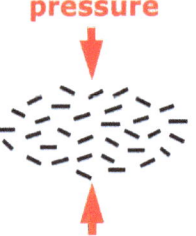 

Randomly oriented crystals can be squeezed flat and into alignment by heat and pressure.

Marble:

Quartzite:

Metamorphic Rocks	
Description	Rock name
foliated & very fine grained, looks much like shale	slate
foliated & fine grained, looks like shiny, wavy slate	phyllite
foliated & medium to coarse grained, typically looks scaly or flaky	schist
foliated & medium to coarse grained, has light and dark banding	gneiss
non-foliated & medium to coarse grained, composed of calcite and effervesces with acid, and will not scratch glass	marble
non-foliated & medium to coarse grained, composed of quartz and does not effervesce with acid, but will scratch glass	quartzite

Lastly, with respect to rocks, **the rock cycle** is an important geological concept. It illustrates that all types of rocks are continually recycled, as any type of rock can potentially be turned into another type of rock under the right circumstances. For example, an igneous rock can be weathered into particles over time, which can then be transported, deposited, and lithified elsewhere to form a sedimentary rock. The sedimentary rock can be buried over time and subjected to intense heat and pressure, thus turning it into a metamorphic rock. Then, the metamorphic rock may be heated to such a degree that it melts. If this magma cools off and solidifies, it then produces a new igneous rock, and the cycle can start all over.

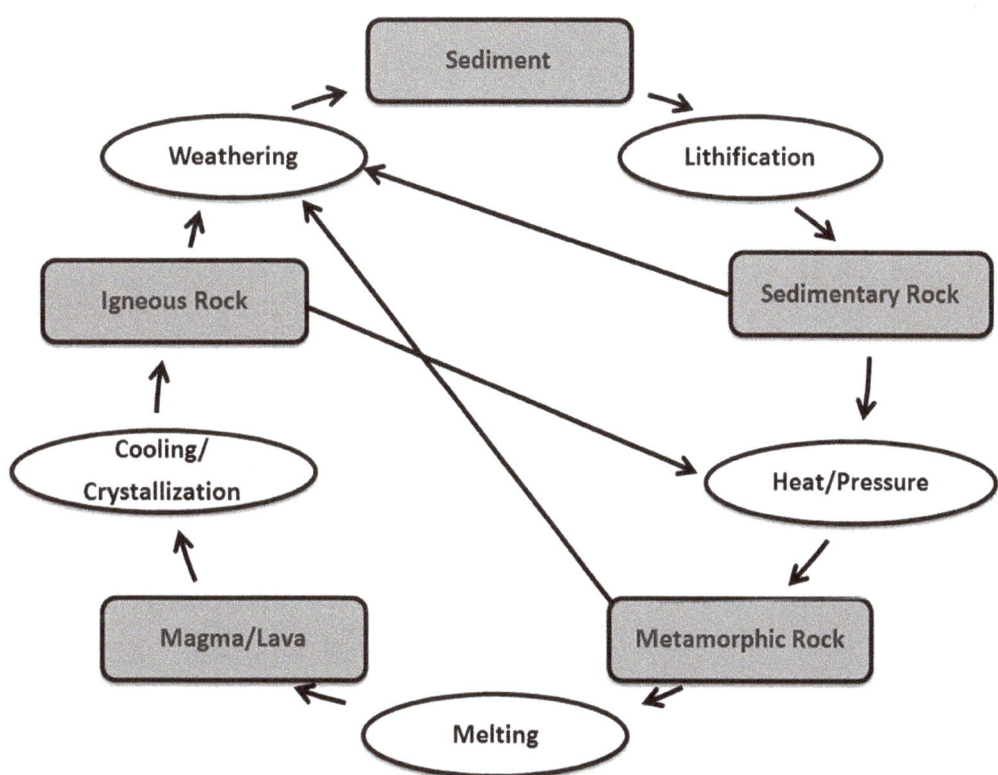

Section 2.2: Weathering, Erosion, and Mass Wasting

Weathering is the mechanical and/or chemical breakdown of rocks. It occurs through a variety of means and produces sediments of all sizes as well as many of the dissolved substances found in water. **Erosion** is the movement of soil, sediment, or rock by wind, water, or ice. The two are distinct processes, but they frequently go hand in hand, because when weathering occurs, erosion often moves the created materials away. **Mass wasting** is the movement of soil, sediment, or rock down slope due to gravity. It may also be tied to weathering, as the moving material is typically created by weathering.

Weathering:

Again, weathering is the mechanical and/or chemical breakdown of rocks. When weathering occurs, it is either through mechanical processes, or through dissolution by water. Thus, it can be considered either **mechanical weathering** or **chemical weathering**, although both frequently occur simultaneously.

A sample of granite that has undergone significant mechanical and chemical weathering. Notice that individual crystals of quartz (colorless and translucent), hornblende (black), and feldspar (light-colored and opaque) are slowly being freed from the rock mass, and that numerous tiny cracks have also developed. These cracks allow water to soak into the rock and increase the rate of disintegration.

There are numerous specific ways that rocks can be broken down by different agents/processes, with the most common being covered here.

Rain:

Rainwater can slowly dissolve some minerals (chemical), and the repeated impacts of raindrops can slowly weaken relatively soft rocks. Above, a sandstone is gradually being turned back into sand (mechanical).

Running water:

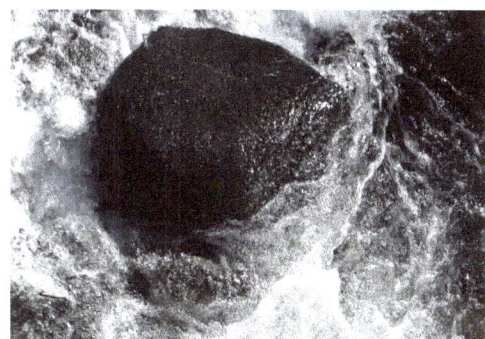 

Water can slowly dissolve some minerals (chemical) and flowing water can carry sediments in suspension, which can repeatedly impact and slowly abrade rocks (mechanical).

Wave activity:

Waves can repeatedly slam into rocks and wear them down, literally beating them apart over time (mechanical). Sediments carried in the water can also abrade rocks (mechanical), and constant exposure to water also promotes dissolution (chemical).

Movement by/in water:

As sediments are rolled along or carried in suspension in water, they repeatedly collide, which slowly wears them smooth (mechanical). Above are rounded gravel on a beach, and gravel being rolled underwater near another.

Wind:

Sand can be picked up and blasted against rocks by strong winds, which abrades them (mechanical). Sandstorms, like the one pictured on the left, are responsible for sculpting the sandstone structure on the right.

Heating and cooling:

Air expands when heated, and contracts when cooled. Rocks do the same, albeit not to the same degree. Repeated expansion and contraction can still cause them to slowly disintegrate over time, though.

Frost/Ice wedging:

 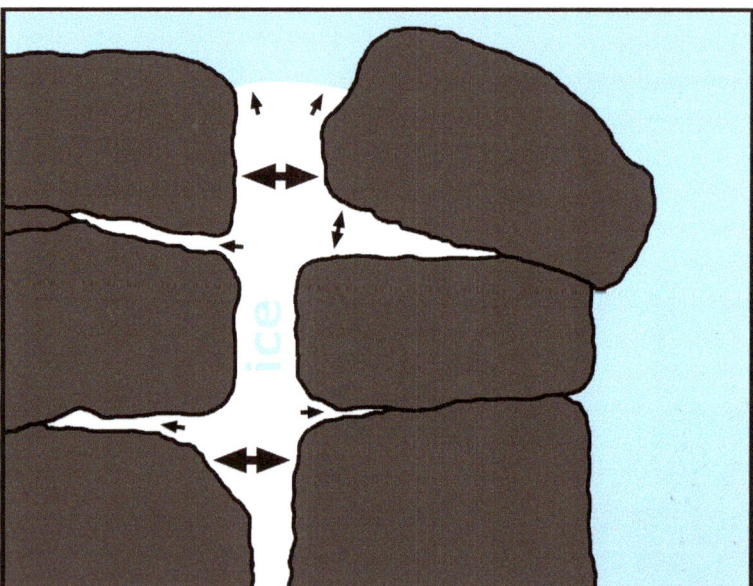

When water freezes and turns to ice, it expands with great force. This can split and push rocks apart (mechanical).

46

Glaciers:

Glaciers carrying sediment can grind rocks they pass over, often leaving scratches/streaks on them (mechanical).

Biological activity:

Lichens and mosses can slowly dissolve rocks with the acids they produce (chemical), while the roots of trees can break rocks apart (mechanical).

Carbonic acid:

When CO_2 dissolves into water, some of it becomes carbonic acid. Then, acidic water can dissolve some minerals/rocks, and especially limestone (chemical). Caverns are formed when acidic water dissolves holes in limestone, which can grow to enormous sizes over long periods of time.

Erosion:

Again, erosion is the movement of soil, sediment, or rock by wind, water, or ice. Frequently, agents of weathering are also agents of erosion, as the breakdown of rocks produces sediments, and those sediments are typically transported some distance away by whatever was responsible for the weathering.

Movement by wind:

A sandstorm in Death Valley, California.

Movement by water:

A fast-moving stream in Denali National Park, Alaska.

Movement by ice:

A large glacier in Prince William Sound, Alaska.

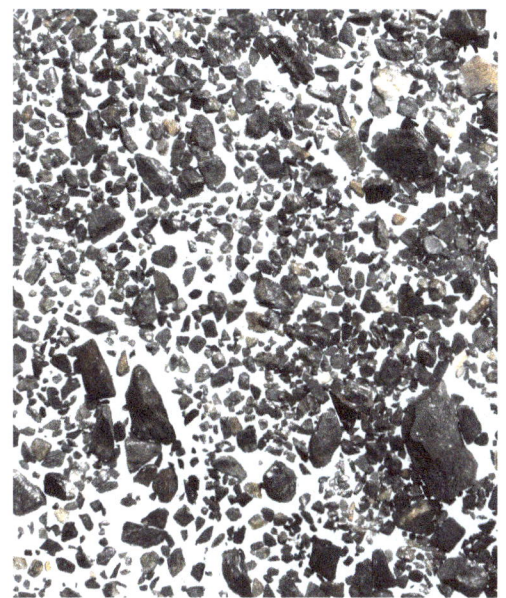

Sediments of all sizes, including boulders, can be carried by glaciers. Above are a piece of sediment-laden ice from a glacier in Alaska, and a boulder left behind by a melted glacier in Sequoia National Park, CA. Seemingly out-of-place boulders such as this are called glacial erratics.

Mass Wasting:

Again, mass wasting is the movement of soil, sediment, or rock down slope due to gravity, which most often occurs as some form of slide or flow, such as a rock slide or a mudflow. Definitions of several types of mass wasting vary, but the common types are covered here with widely-accepted definitions.

A **rockfall** is the downward movement of one or more relatively large rocks that breaks away from an exposed rock face, by free-fall, rolling, bounding, or rapid sliding (or a combination of these). A **rockslide** is the downward movement of previously accumulated rock rubble as a unit, primarily by rapid sliding. And, a **rock avalanche** is the downward movement of previously accumulated rock rubble as loose and fast-moving material, primarily by rolling or bounding.

Rockfall boulders in Iceland.   A rockslide in Alaska.

A **landslide** is the downward movement of loosely consolidated soil, sediment, or rock that slides a relatively short distance. These are called a **slump** when they occur on a curved surface.

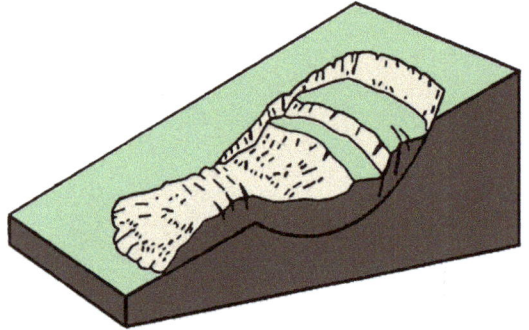

The term landslide is often used for a number of mass wasting events. However, it's proper use is when there is a distinct zone of weakness that separates the slide material from more stable underlying material. Also, a landslide is properly called a slump only when material moves along a curved surface of detachment from the underlying material. Above is a diagram of a slump, with a large one pictured on the right.

A **debris flow** is the rapid downward movement of unconsolidated, loose material that is wet, with over half of the moving sediments being sand-size or larger.

Lastly, a **mudflow**, which is often called a mudslide, is the rapid downward movement of unconsolidated, loose material that is wet, with over half of the moving sediments being sand-size or smaller.

A simple diagram of a debris flow or mudflow. The difference in the two is the size of the moving sediments.

So, why do these differing forms of mass wasting occur? Of course, there are a variety of reasons. The detachment and subsequent movement of rocks/sediments may be the result of:

Rocks becoming detached due to frost wedging, the intrusion of tree roots, or even earthquakes.

Rivers and streams undercutting a slope as they erode material.

Changes in vegetation and a loss of root systems, which help hold sediment in place.

Overloading caused by human construction.

Tree roots can breaks rocks apart, but can also hold sediments together, as can the roots of smaller plants.

Or, an **increased water content** in materials after heavy rainfalls or snowmelt. Water is particularly influential, as it can act as a lubricant between grains of sediment. This reduces the friction between them, which helps hold them together and in place.

Section 2.3: Surface Water and Groundwater

When rain reaches the Earth's surface, or snow and ice melt, some of the water may evaporate, but most will either flow downwards over the surface or soak into the ground under the influence of gravity. Appropriately, that which moves over the surface is called **surface water**, while that which soaks into the ground in called **groundwater**. Surface water typically flows in streams of various sizes until it reaches a lake, an ocean, etc., while groundwater may accumulate in subsurface areas called aquifers. Both can create a variety of geologic features.

Surface Water:

Again, surface water tends to move from land to the sea in streams, with a **stream** being any channelized flow of water, regardless of its size. So, this means that creeks, brooks, streams, and rivers are all streams by definition, just streams that vary in size.

Streams play a significant role in the shaping of landscapes, as they can weather rocks, and erode and deposit sediments. They also move huge amounts of material from continents to the oceans.

Streams transport material three ways:

The majority of material is carried in:

Rivers typically look "muddy" because they carry large amounts of fine sediment in suspension.

Again, streams can come in any size, but rivers are the focus here. Rivers can start, run, and end in various ways depending on the geology, topography, and climate of an area, but the majority start in elevated areas called **headlands** and end in the sea, with the end of a river where it meets the sea (or other body of water) being called its **mouth.** So, a river starts in headlands and ends at its mouth.

In headlands, small streams typically form a branching **dendritic drainage pattern**, which looks much like the roots of a tree. Small streams join to form larger ones, and those join to form even larger ones, etc. These relatively small streams are called **tributary streams**, as each contributes water to the river.

In headlands, large amounts of sediment can also be **eroded** and transported downstream, as the water often flows quickly down relatively steep slopes.

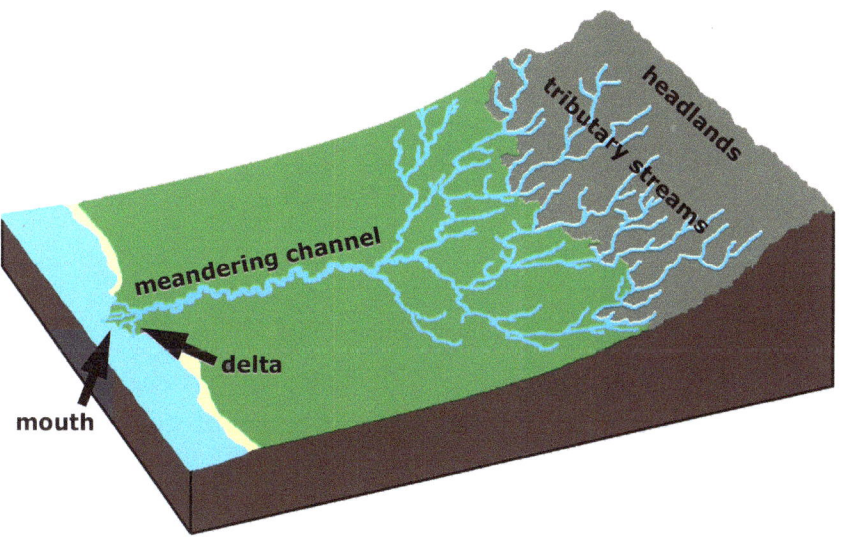

Eventually, tributary streams feed into a larger, and repeatedly curving **meandering channel**, which is the river itself. In this part of a river system, **both erosion and deposition occur**, as water flows at different velocities in different parts of the channel.

Where a channel is relatively straight, it typically has a **symmetrical cross-section**, with the highest flow velocity being in the middle and near the surface. However, in the curve of a meander, a channel typically has an **asymmetrical cross-section**, and the area of highest flow velocity moves to the outside of the curve.

Above, the headlands of a river seen from space, with a well-developed dendritic drainage pattern. Below, a typical meandering channel.

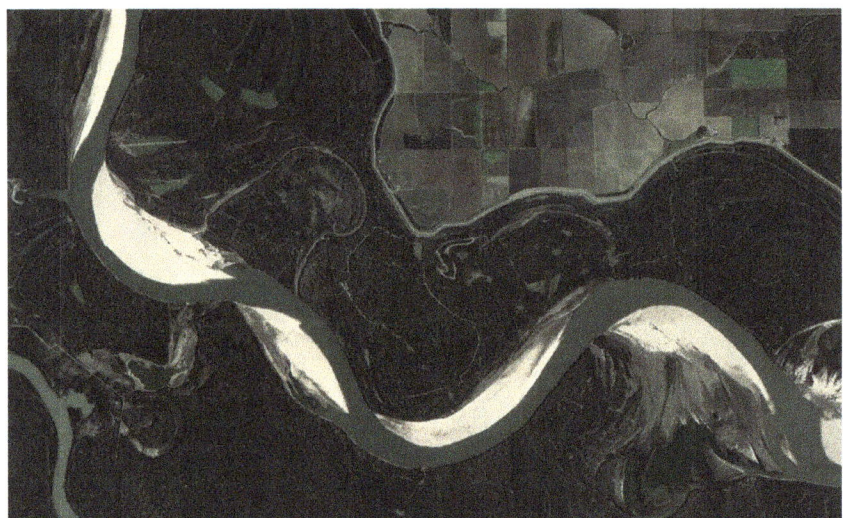

In a meander, the more the channel curves, the closer this area of higher flow moves to the outer channel bank. Conversely, flow velocity decreases near the inside, along the inner bank. For this reason, sediment is typically **eroded away from the outside** of the curve, and is simultaneously **deposited on the inside**. This leads to the development of a **cut bank on the outside** and **point bar on the inside** of meanders. Note that any such sediment deposited by a stream is called **alluvium**.

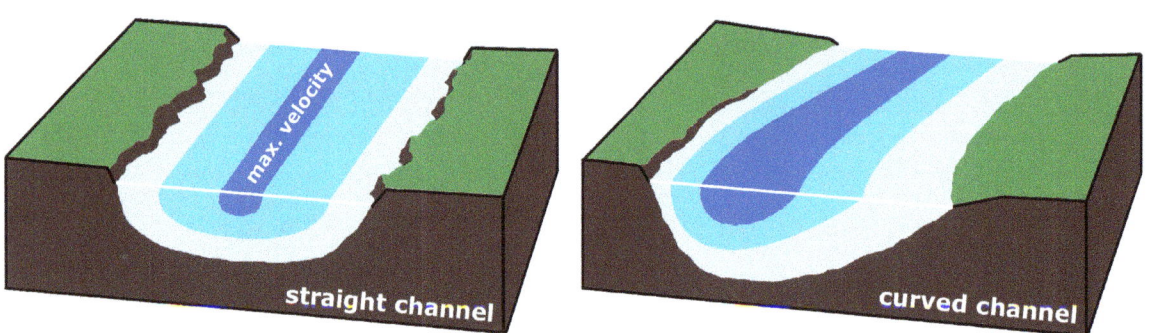

Channels are typically symmetrical in cross-section where straight, and asymmetrical in cross-section where curved. Erosion will take place on the outside of a meander where the channel bottom is steeper and flow velocity is higher, while deposition will occur on the inside of a meander where the channel bottom is less steep and flow velocity is lower.

Again, when a river meanders, erosion occurs on the outside bank of each meander and deposition occurs on the inside bank of each. So, the meanders, and thus the river itself, will move laterally over time. As this occurs, several features can be produced.

Cut bank:

Point bar:

Meander scar:

Abandoned meander:

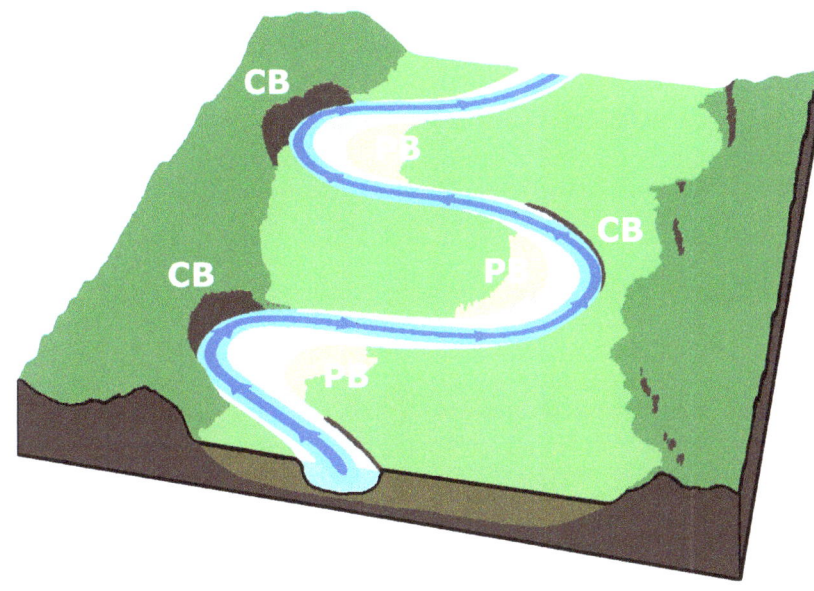

A meandering channel creates cut banks (CB) and point bars (PB) along its course. Note that the area of flow with the highest velocity shifts to the outside of each meander, which erodes material from the outer bank.

Oxbow lake:

Floodplain:

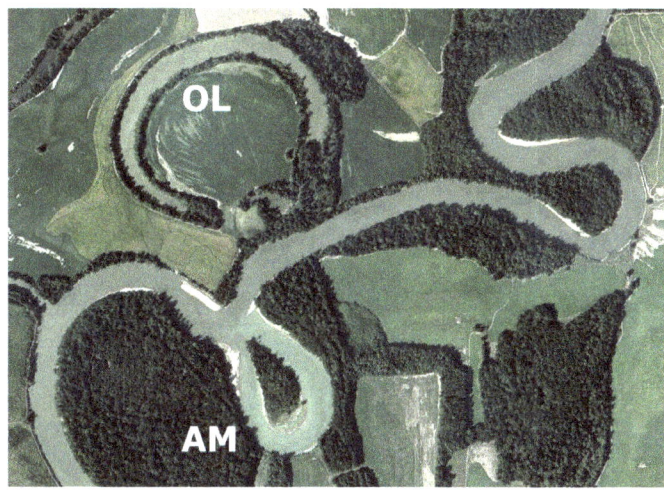

An oxbow lake (OL) and an abandoned meander (AM).

An oxbow lake can be created when two meanders abandon another. When this occurs, water flow though the abandoned meander effectively falls to zero, and it begins to fill with sediment dropping out of suspension. Eventually the "ends" of the abandoned meander can be closed off from the river, and it then becomes an oxbow lake.

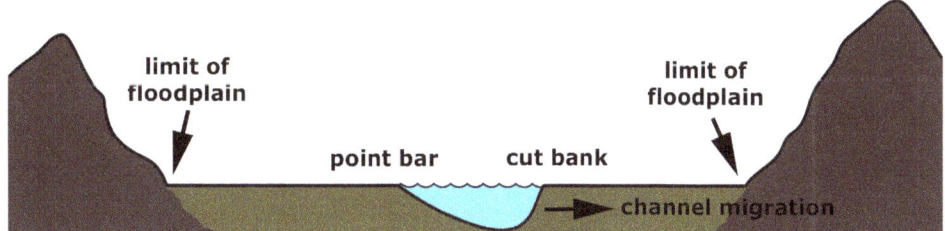

Over time, the lateral migration of a channel can create a floodplain.

As a river continues to migrate laterally over time, a floodplain can grow and expand its limits.

However, in some situations the geology and/or topography of an area may hinder migration, and prevent the formation of a floodplain at all. A river that flows through a steep-sided rocky canyon and maintains its course over long periods of time is a good example, such as where the Colorado River flows through Grand Canyon.

Eventually, when a river reaches the sea, it is at its **base level**, which is the lowest point that it flows to. Then, the water typically fans out, and its velocity falls precipitously. As this occurs, large amounts of sediment are deposited, thus creating a **delta**, which is a large deposit of alluvium at the mouth of a river.

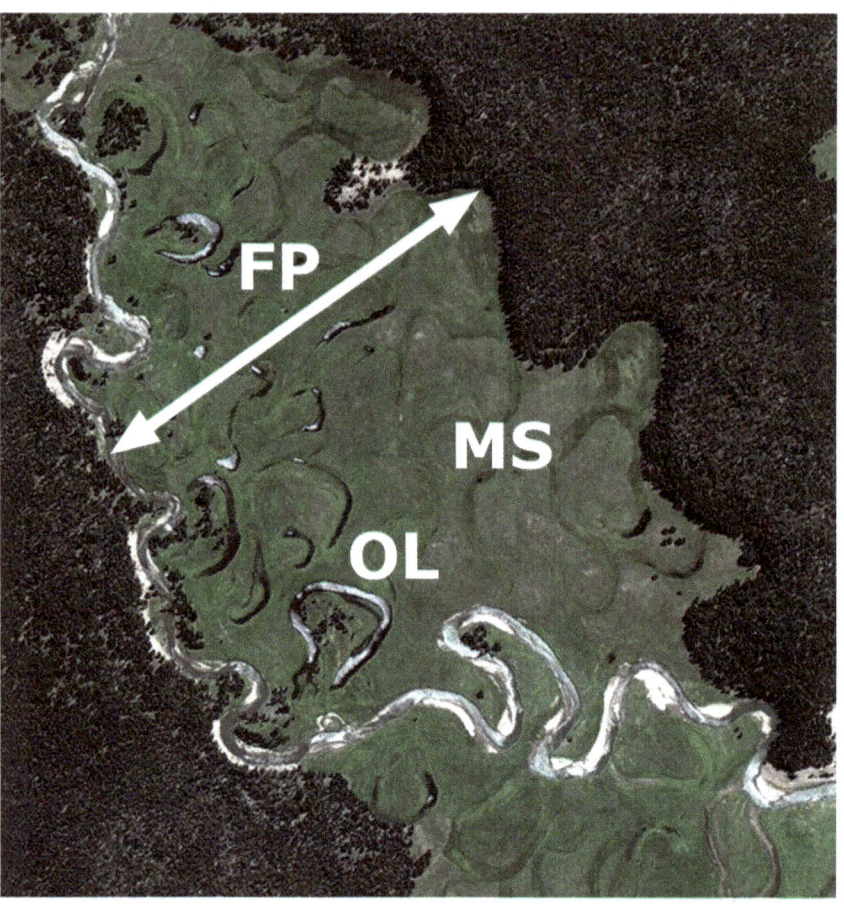

This river has created a well-developed floodplain (FP). Meander scars (MS) and a few oxbow lakes (OL) can also be seen.

The Nile River delta, which has a distinctive triangle shape, like the Greek letter delta (Δ).

The Mississippi River delta, which is called a "birdfoot delta" due to its odd shape.

Groundwater:

If water soaks into the ground, it will move downward under the influence of gravity until it eventually reaches material that it cannot penetrate. At times, impenetrable material may be at the surface, preventing the water from moving downward in the first place, but typically at least some water can move through materials, albeit very slowly. However, it may easily move through some materials.

How well water can move though sediments and rocks depends on two primary factors, which are the material's **porosity** and **permeability**.

Porosity:

Permeability:

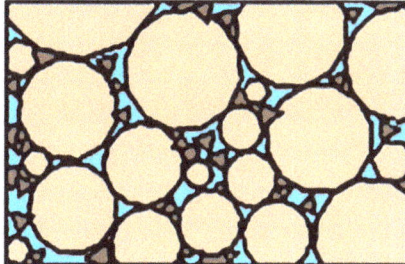 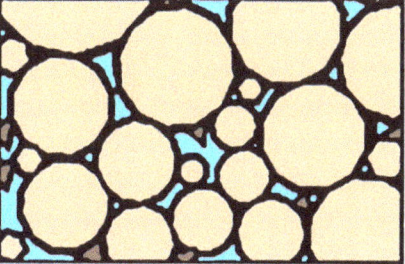 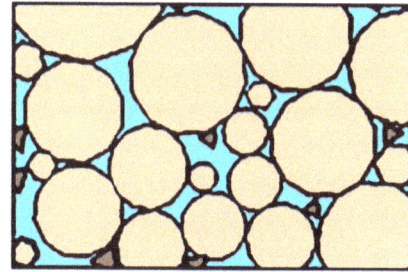

The material on the left has relatively low porosity due to the presence of many fine particles in between grains of sand. That in the middle has better porosity (fewer fine particles present), but poor permeability due to the presence of cement around the grains. And on the right, material that has good porosity and permeability due to the low number of fine particles and absence of cement.

A material having very low porosity and/or permeability, which does not allow water to flow through it is called either an **aquitard** or an **aquiclude**.

Aquitard:

Aquiclude:

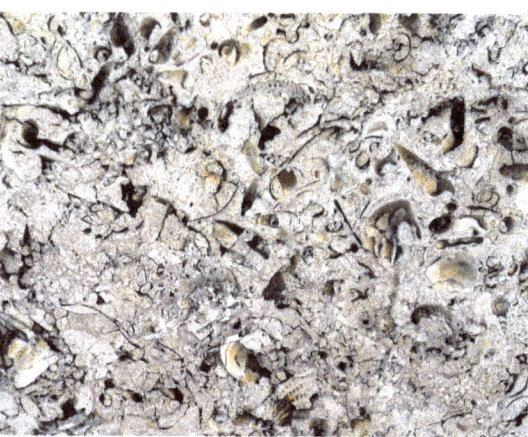

Porous Florida limestone.

Once water reaches an aquitard or aquiclude, it will begin to accumulate in the pore spaces in the material above it. This area, where all pore spaces are filled with water, is the **zone of saturation**, and the top of the zone is the **water table**. **Groundwater** is all the water in this zone.

To the right, water poured into a glass of sand can be used as an analogy. When poured in, the water soaked through the sand until it hit the bottom of the glass. So, the bottom of the glass acted as an aquiclude.

Then, the water began to fill all the pore spaces in the sand at the bottom of the glass, and rise. Thus, the bottom half of the glass, where all the pore spaces are visibly filled, is analogous to the zone of saturation (ZOS).

All of the water in the bottom half of the glass would be groundwater (GW). And, the top of the zone of saturation, where the sand becomes visibly dry, would be the water table (WT).

At times, a relatively small layer of impermeable material may also create a **perched water table**, one that develops above the primary zone of saturation in an area. Regardless, if the water table, or a perched water table, meets the surface and water is able to flow out from the zone of saturation, a **spring** is formed, which is a naturally occurring flow of groundwater from the ground.

Oftentimes, streams and other bodies of water at the surface are fed by water flowing from the zone of saturation, as well. If a stream receives most of its water from the ground, rather than from precipitation, it is called a **spring-fed stream** or **spring-fed river**. These are common in Florida.

Also, if a subsurface area stores a significant amount of groundwater, it is called an **aquifer**, and any hole dug or drilled from the surface to an aquifer is a **well**.

To continue with some additional groundwater-related features, a **hot spring** is a spring producing water that has been heated by hot rock or magma. And, a **geyser** is a hot spring that periodically ejects a fountain of boiling water and steam into the air.

A small mountain spring in Alaska.

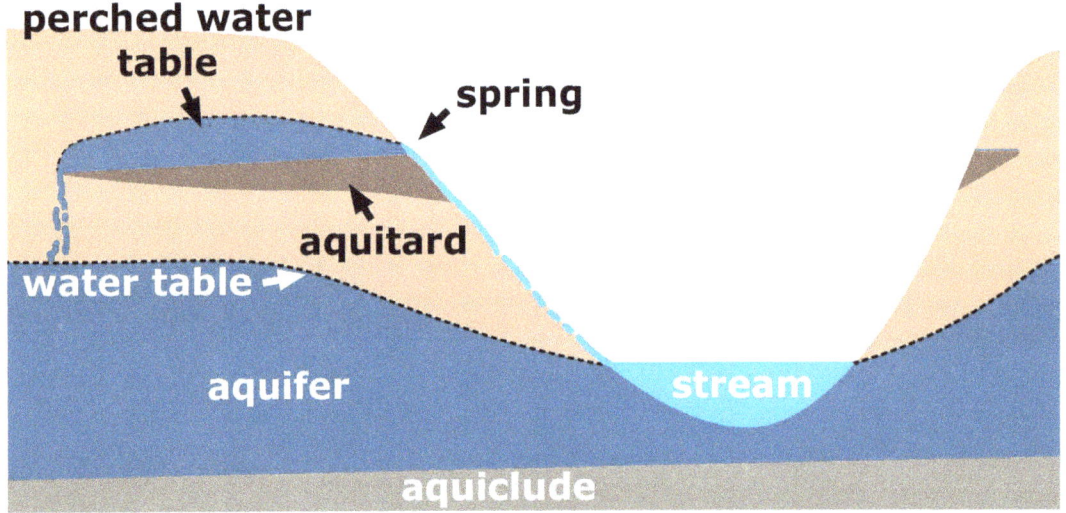
Several groundwater-related features, including a spring and spring-fed stream.

Above is a small hot spring in Yellowstone National Park, Wyoming. The colors are actually different species of bacteria that live in the spring. On the right is Old Faithful, the most famous geyser in the world, which is also in Yellowstone N.P. It has been "faithfully" erupting every 35 to 120 minutes since its discovery in 1870.

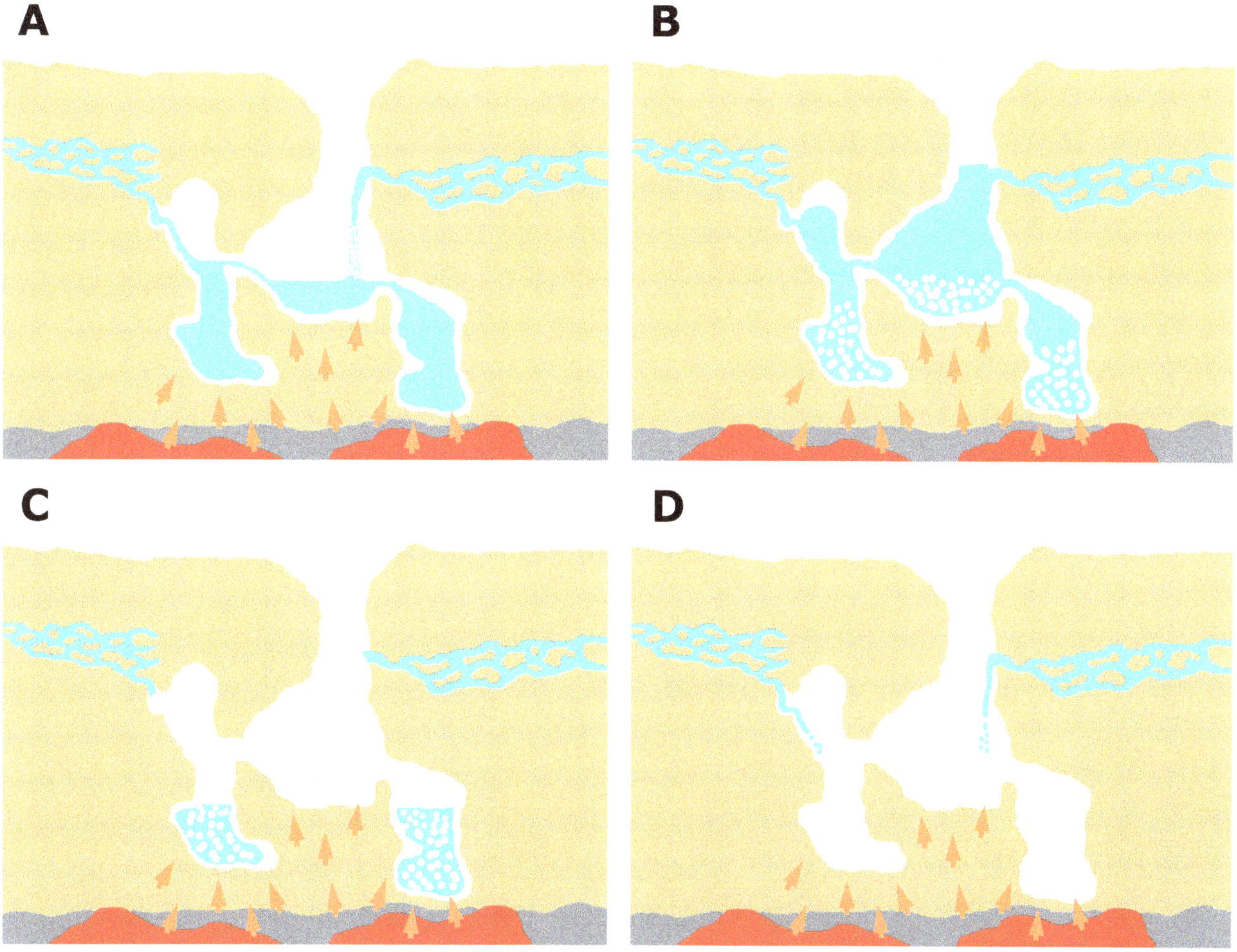

A

B

C

D

Geysers only form in particular geologic settings, where caverns are present, groundwater is plentiful, and heat from nearby magma is sufficient to boil water. The basic stages are: A) A cavern with an opening at the surface fills with groundwater. B) Heat from underlying magma eventually boils the water. C) The geyser erupts and spews forth essentially all of the water that was in the cavern. D) The cavern begins to refill and restart the process.

Caverns, subsurface chambers formed by the dissolution of limestone, are also formed by groundwater. As covered earlier, water may become acidic if CO_2 is added and forms **carbonic acid**. Some of the CO_2 comes from the atmosphere, but soil can also contain significant amounts of CO_2, which can make water soaking through it and into underlying rocks even more acidic.

Limestone (calcite) can be dissolved by carbonic acid. So, as acidic groundwater moves through limestone, it can slowly dissolve it away. Initially all caverns start as tiny holes produced this way, but over time they may grow to enormous sizes. Note that the growth of a cavern only continues as long as there is groundwater present, though.

Over time, land can be uplifted, the climate can change, etc., and water tables may thus rise and fall. If a water table should drop enough for a cavern to become partially or completely filled with air, a number of **speleothems** may form. These are features in caverns that form through the precipitation of minerals, rather than by dissolution.

As acidic water dissolves limestone, the released calcium is carried in the water. Then, if this calcium-laden water drips from the ceiling of a cavern or trickles down its walls, some of the CO_2 is re-

Speleothems in Cave of the Winds, CO.

leased, and some of the calcium is turned back into limestone (calcite). At first this makes small tube-like features on the ceiling, often called **soda straws**. However, these may grow to large sizes over time and are then called **stalactites**, which are elongated speleothems that grow from the ceiling of a cavern.

If the water table falls below the floor of a cavern, and the floor becomes relatively dry, water may also drip off the ends of stalactites and splash on the floor. This can release more CO_2 from the dripping water, and thus cause the precipitation of more calcite. This building up of rock from the floor leads to the formation of **stalagmites**, which are elongated speleothems that grow from the floor of a cavern.

As stalactites and stalagmites grow toward each other, they may eventually meet and form a **column**, and all of these limestone features made by the dripping of water are collectively called **dripstones**. In addition to these, as some water trickles down the walls of a cavern, **flowstone** can be produced on them through the same degassing process.

The beginnings of dripstones.

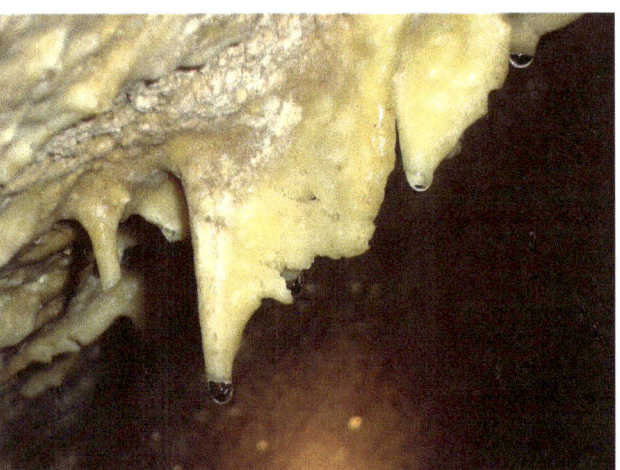

Growing dripstones.

Stalactites, stalagmites, and columns in Carlsbad Caverns, New Mexico.

Stalactites, a stalagmite, and a column in Cave of the Winds.

Flowstone in part of Mammoth Cave National Park, Kentucky.

Lastly, at times a cavern can grow too large and cannot support the weight of overhead material. When this occurs, the cavern may cave in. This produces a **sinkhole**, which is a natural depression formed by the collapse of a cavern's ceiling. Florida is composed primarily of limestone and thus has many caverns, spring-fed streams, and sinkholes, and areas around the world, including Florida, are called **Karst regions** if they have these characteristics. Of course, the formation of sinkholes can be promoted by human activities when the construction of infrastructure and buildings takes place in such areas.

A typical sinkhole in Winter Park, FL.

A water-filled sinkhole on San Salvador Island, Bahamas.

These lakes around Winter Park are water-filled sinkholes, as are most of the ponds and lakes in Florida. Sinkholes, large and small, are exceptionally common in the central part of the state.

Section 2.4: Glaciers

A **glacier** is a mass of ice formed by the accumulation, compaction, and recrystallization of snow. They form in cold areas where snow accumulates year after year, near the poles, and on tall mountains in other areas. As snow accumulates in these areas, it is increasingly compacted by the weight of new snow, and this essentially squeezes the underlying snow into solid ice. Then, despite being solid ice, they slowly flow downward due to gravity and eventually melt at lower/warmer elevations, or end in a large body of water.

In polar regions, it is cold enough for snow and ice to remain all year, and many mountains have a **snow line**, which is the point above which it's cold enough for snow and ice to cover the ground throughout the year. Such an area, where snow and ice persist all year, is a **zone of accumulation**. Also, in the case of a glacier moving to lower/warmer altitudes and melting, the area where melting is greater than accumulation is the **zone of ablation**.

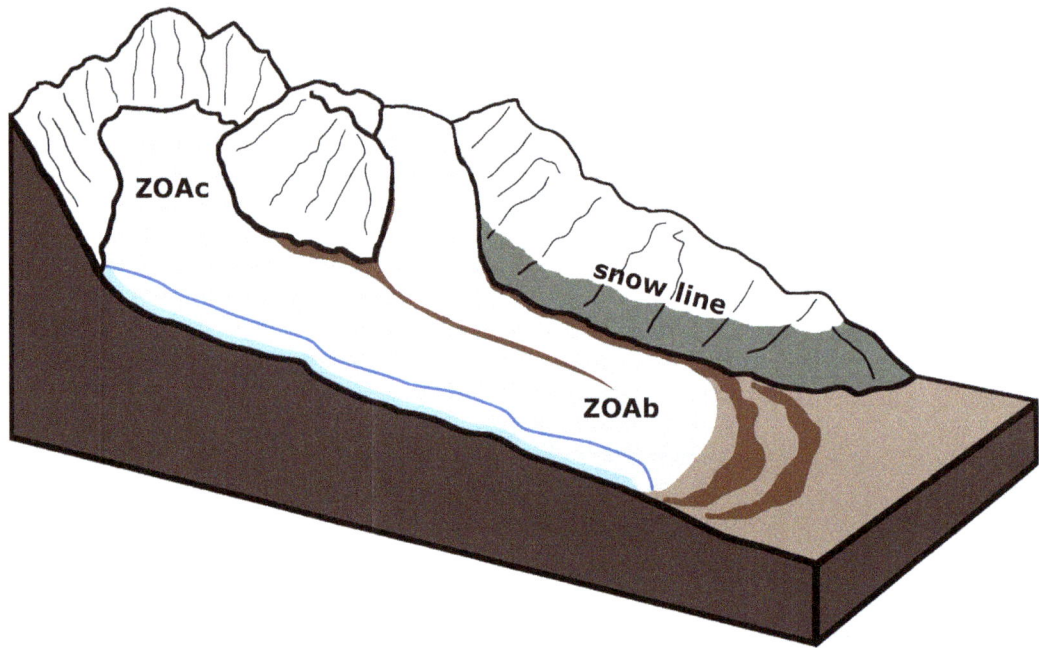

A glacier, with the snow line, zone of accumulation (ZOAc), and zone of ablation (ZOAb) shown.

Those in mountains are called **alpine** or **valley glaciers**, but they may reach the base of a mountain and spread out over relatively flat land there. They're then called **piedmont glaciers**. Others may end in a lake or in the sea, and are then called **tidewater glaciers**. Tidewater glaciers are the source of **icebergs**, which are created as pieces of ice break, or calve, off the end of the glacier and float away.

Icebergs calve from the ends of tidewater glaciers and are especially deceptive - and dangerous - because about 90% of their mass is below the surface and unseen.

An alpine glacier (top), piedmont glacier (middle), and tidewater glacier (bottom), in Alaska.

As noted earlier, glaciers flow to some degree, despite being made of solid ice. This is due to pressure. When ice, which is typically a brittle solid, is placed under sufficient pressure it will soften and act as a **plastic**, which is an easily-deformable solid. This requires significant pressure though, so **the surface of a glacier will remain brittle**, while **only the ice deeper than approximately 150 feet will soften**. For this reason, large cracks called **crevasses** form in the upper part of a glacier as it moves over any rough terrain, around curves, etc. Ice melts more easily when under great pressure, so liquid water may form at the base of thick glaciers, which can act as a lubricant and aid in their downward flow, as well.

Numerous large crevasses have developed in this tidewater glacier due to its passing over uneven surfaces.

Glaciers can weather rocks, erode materials, and deposit them, too. As a glacier moves over the surface, some sediments are picked up, or plucked, and incorporated into the ice. Sediments may also accumulate on top of a glacier due to the weathering of rocks above it, primarily by frost wedging, and also be moved away. Significant weathering of any rocks a glacier moves over can also occur, as they are abraded by the plucked material. And, the plucked material and any that was on top will be dumped wherever a glacier melts. Any such material deposited by a glacier is called **glacial till**.

Plucked sediments (left) are incorporated into the bottom of a glacier and can then abrade other rocks.

Glacial till can range in size from the finest mud to building-size boulders, and is typically poorly sorted.

Glaciers typically form in mountains where bowl-like depressions called **cirques** develop. From there, they move downward, and as they weather and erode the landscape, **glacial troughs** are created, which are U-shaped valleys. U-shaped **hanging valleys** may also be created and left behind where a large glacier was fed by smaller **tributary glaciers**, but later melted away. The bottom of a hanging valley is typically higher in elevation than that of the main trough, where the larger glacier they fed was present. Also, sharp peaks called **horns** are commonly created where three or more cirques meet, and sharp ridges called **arêtes** are often formed where two troughs are close to and parallel to each other.

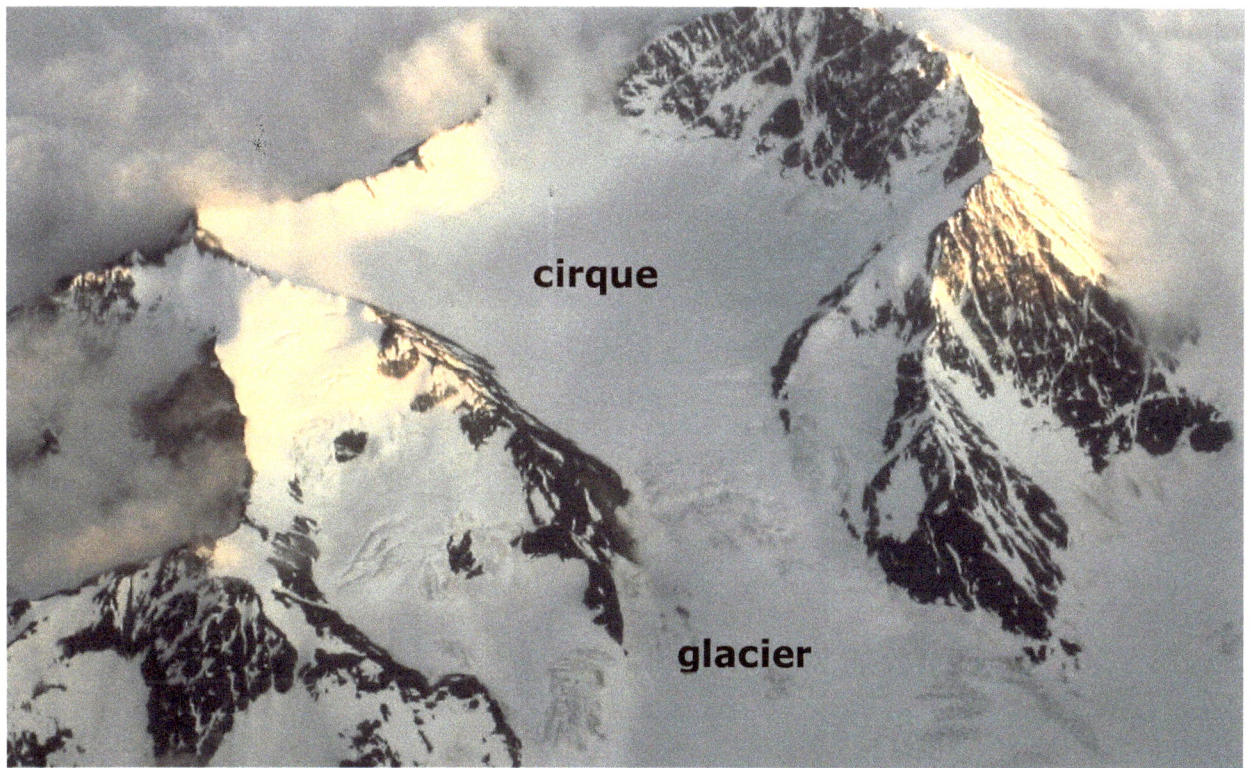
A large cirque, feeding snow/ice to a glacier moving downwards from it.

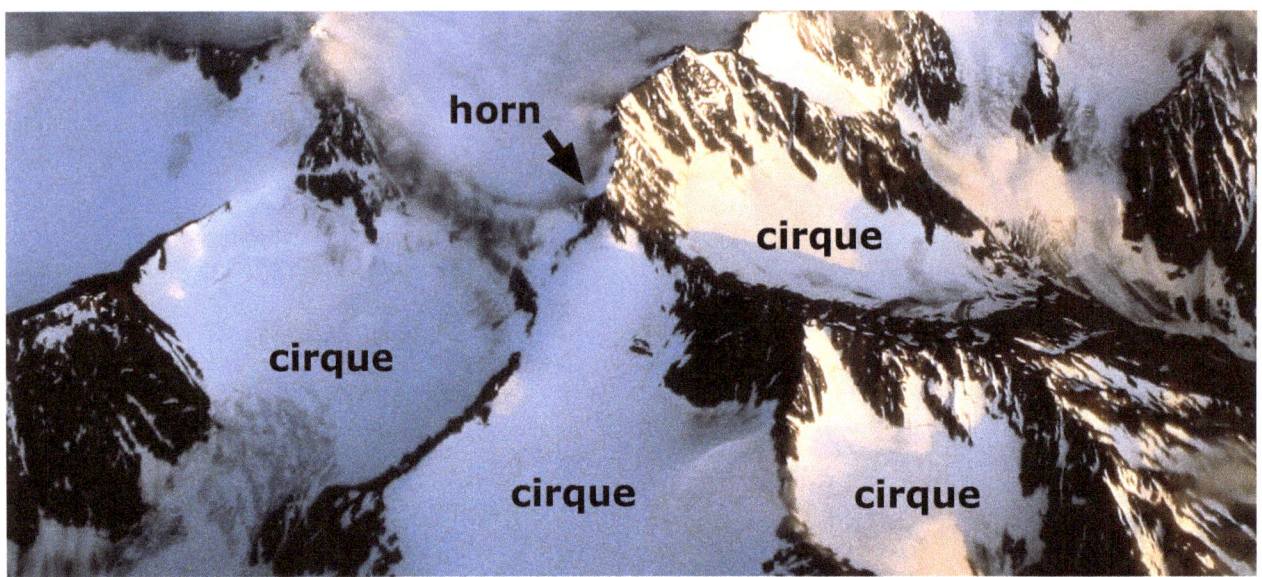

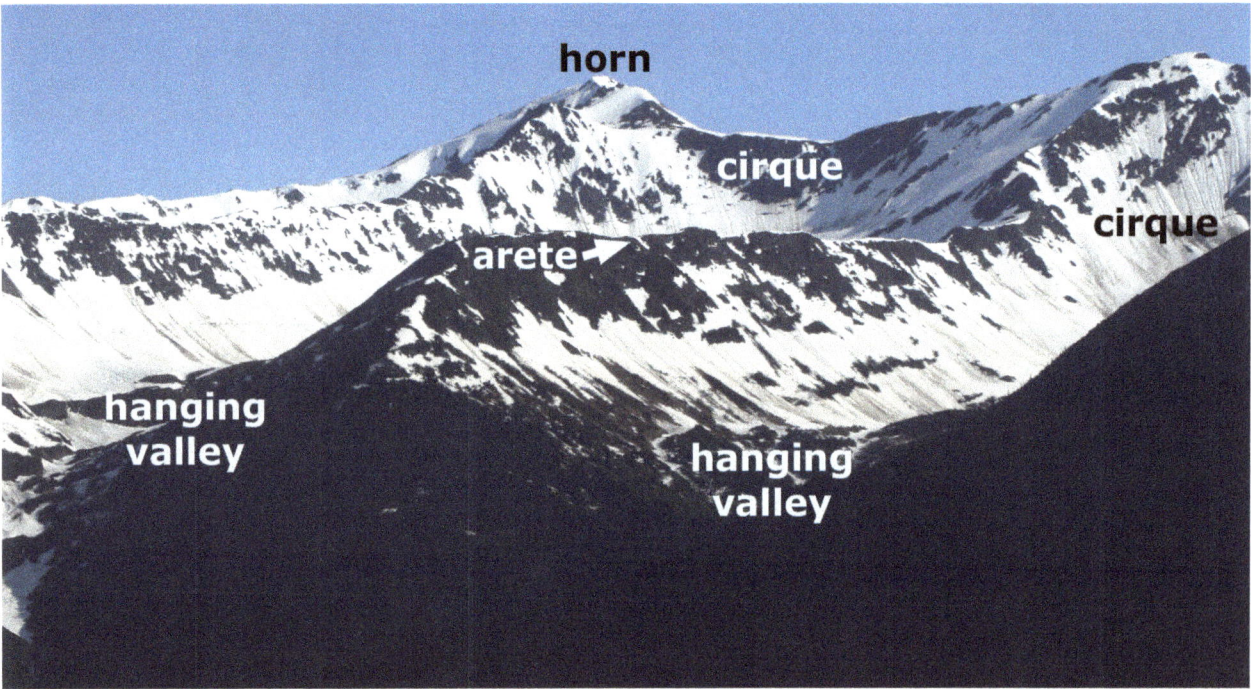

Cirques, horns, arêtes, and hanging valleys in Alaska. These were occupied by significant glaciers in the past.

Also, as weathered material accumulates along the edges of glaciers, a **lateral moraine** is created upon it. And, when two glaciers meet, these may come together to form a **medial moraine**. These are the stripes of material often seen running down the middle of larger glaciers. At a glacier's end, where large amounts of material are deposited in the form of till, a **terminal moraine** can be formed, as well.

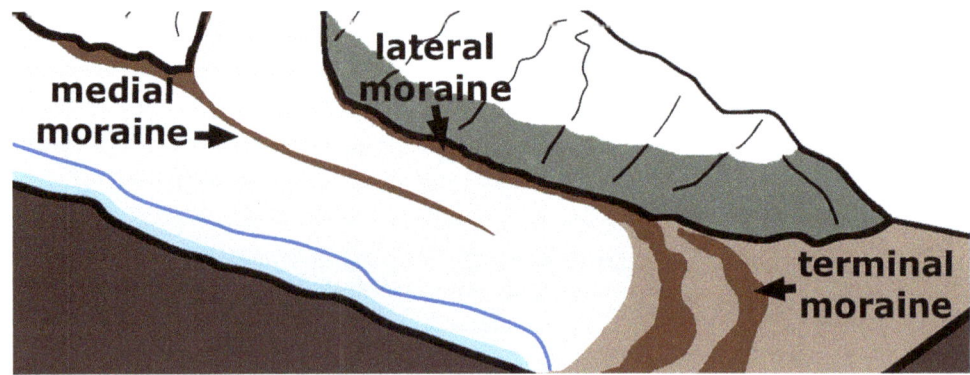

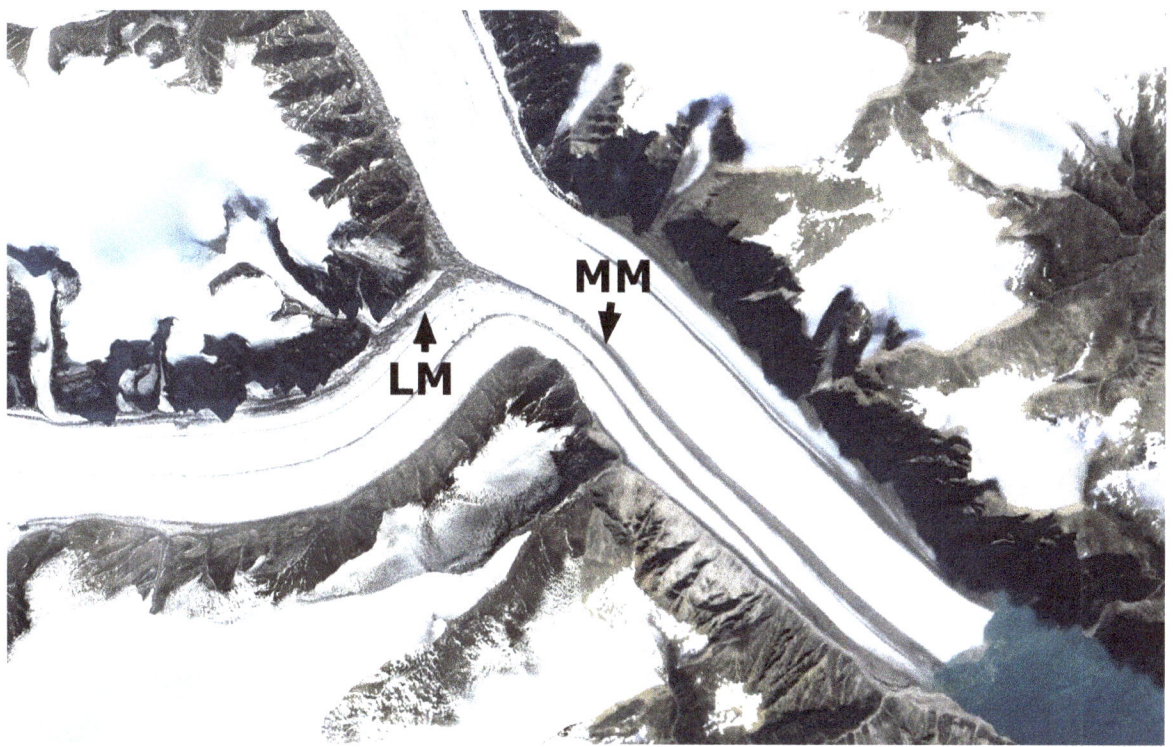

Lateral moraines (LM) can join to form medial moraines (MM) where two glaciers meet. If more than one medial moraine is present on one glacier, as above, then more than two smaller glaciers joined to form it.

As will be explained later, sea level changes significantly as the Earth's overall climate changes. So, **fiords**, which are glacial troughs that have been flooded by seawater, can be formed in some places.

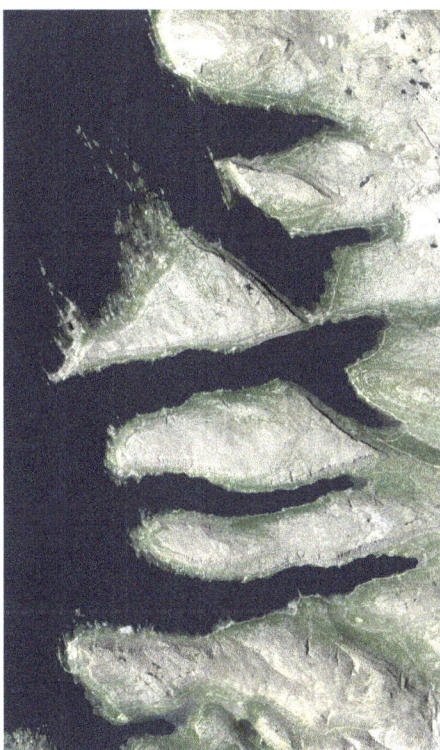

Several fiords, absent of glaciers.

Two joined fiords filled with icebergs from tidewater glaciers.

In addition to these glaciers, some areas are covered by relatively huge **ice sheets**, which are thick masses of glacial ice covering more than 20,000 square miles of land. These also flow downward, but more so in all directions as snow accumulates upon them. In the past, much more land was covered with such sheets, but on today's much warmer Earth, there are only two. These are the ice sheets covering Greenland and Antarctica, which are approximately **10,000** and **14,000 feet thick**, respectively. Thus, they hold a tremendous amount of water in a frozen state.

Greenland

Antarctica

Lastly, the Earth does warm up and cool off over millennia, which is why there is more or less ice on its surface at different times. When the Earth cools, it moves into an ice age, which is properly called a **glacial period**, and there have been many in the past. On the other hand, when the Earth warms it moves into an **interglacial period**. These swings in global climate are caused by three cyclic changes in Earth's orbit and orientation, collectively called **Milankovitch Cycles**.

Earth's orbit is not a perfect circle, and neither are the orbits of any of the other planets, or the moons going around them. It also changes in shape, slightly, over thousands of years, moving back and forth from being a little closer to a perfect circle to a little farther from one. Earth's axis is now tilted approximately 23.5°, but that changes slightly over thousands of years too, as it leans over a little more, then a little less. And, the direction 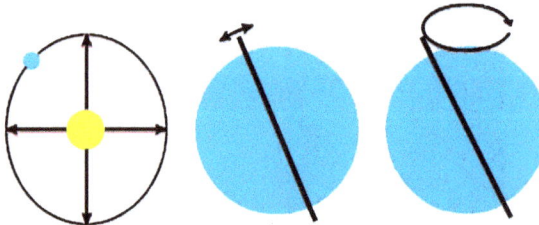 that Earth's axis points in space varies over thousands of year, as well. All of these changes affect Earth's climate, and can drive temperatures up or down, again, over many thousands of years.

These climactic changes not only affect the amount of ice on Earth's surface, but also affect sea level. When the Earth cools, the amount of ice on land increases greatly because as water evaporates from the oceans, more of it snows on land and is then stored there in the form of glaciers and ice sheets. Thus, sea level goes down. Conversely, when the Earth warms, this ice on land begins to melt and the water flows back to the oceans.

The difference is quite large, as sea level at the peak of the last glacial period was **about 400 feet lower** than it is today, and approximately 20 feet higher at the peak of the last interglacial period, which did not significantly melt the Greenland or Antarctic ice sheets. All glacial/interglacial periods are not the same, though. During an interglacial period warm enough to melt the two existing ice sheets away completely, sea level would **rise about 260 feet**, and submerge vast areas of land.

Section 2.5: Deserts

Deserts, which are also called arid regions, are areas that **receive less than 200mm of precipitation per year**, and are not considered to have a polar climate. To make a comparison, 200mm of precipitation per year is about 8 inches, the U.S. average is 37 inches per year, and Tampa's is 47 inches per year. Some deserts do get quite cold, but an area with a polar climate is one where the average monthly temperatures do not rise above 10°C, while a desert must have at least one month per year with an average temperature above this threshold. Regardless, precipitation in a desert is so low that they typically have little to no vegetation.

Part of the largest desert on Earth, the Sahara.

Because deserts can have lower or higher average temperatures, they're split into cold deserts and hot deserts. Cold deserts may actually have a warm summer, but they always have very cold and very dry winters. They also tend to be found at higher altitudes. Conversely, hot deserts are **typically found under/near the Subtropical High** where dry air descends (at around 30° north and south latitude), and thus tend to have clear skies, little precipitation, and exceptionally high average temperatures for at least part of the year.

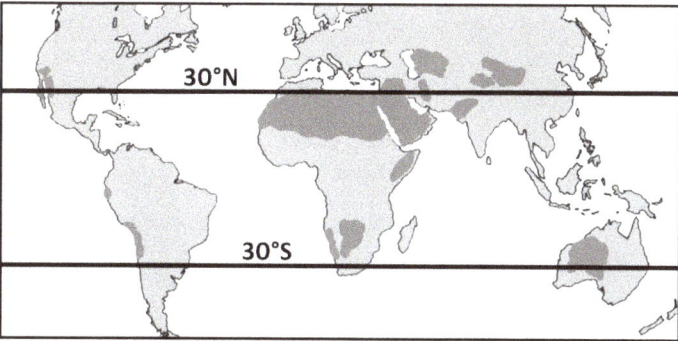
The world's major deserts are indicated with dark gray.

The American Southwest is the nearest desert area, with several named deserts in the region. Thus, it will be the focus here. The major deserts are shown on the right, with the Great Basin Desert being classified as a cold desert, and the rest being hot deserts.

Much of this area is mountainous, and many streams cannot make their way to the sea. Precipitation is relatively low, and evaporation is high, as well. So, many streams and other bodies of water are intermittent or ephemeral (temporary), as they may dry up quickly. Winds can be strong, and fields of sand dunes are scattered about, too. Thus, many notable features are found in these areas, especially in Death Valley National Park, California.

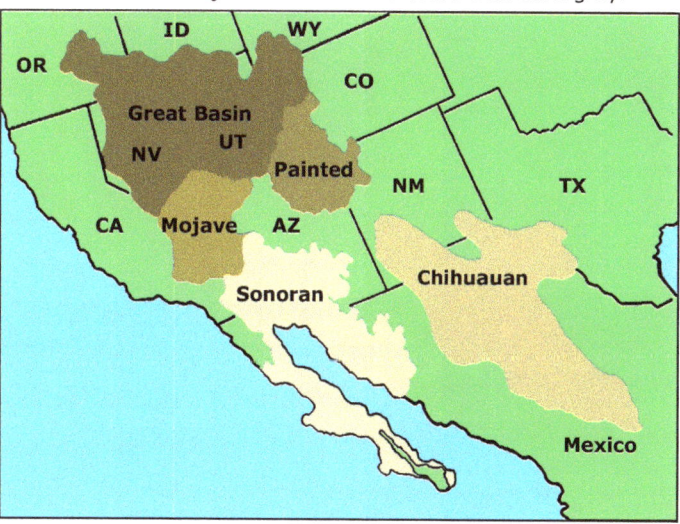
Major deserts of the American Southwest.

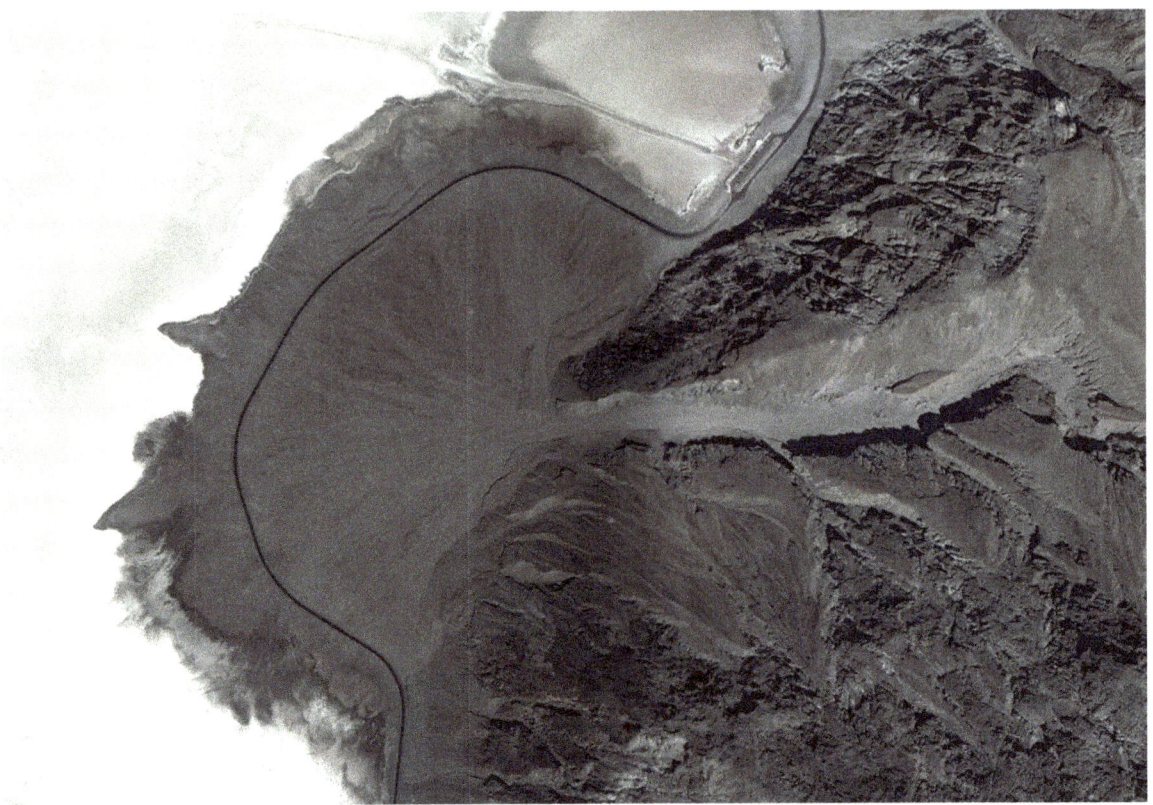

Alluvial fan: Death Valley National Park

Ephemeral stream: DVNP

Ephemeral stream bed:

Intermittent stream:

Intermittent stream bed:

Playa lake: DVNP

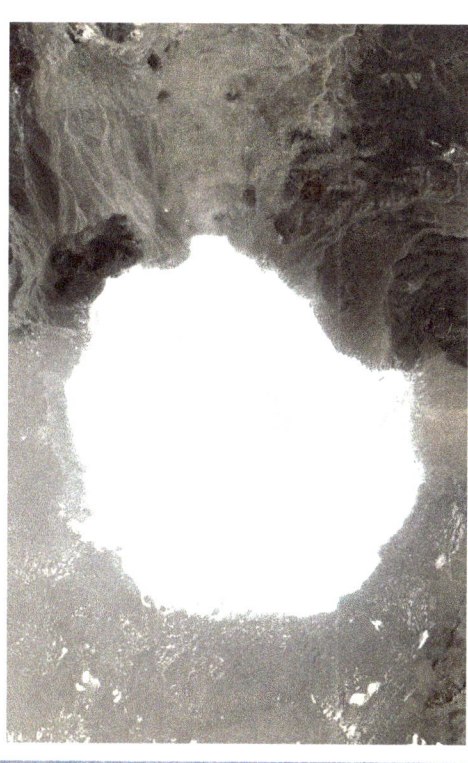

Playa: DVNP

Sand dune:

DVNP

Well-developed dunes are typically asymmetrical in form when wind generally blows in the same direction for an extended time, having a lower slope **windward face**, and a steeper **slip face** on the downwind side. As air flows up the windward face, it can move sand upwards, and with sufficient wind velocity, sand slides along and saltates (bounces) up the windward face towards the dune crest. Then, as the air passes the dune crest, some sand will settle and accumulate there. Eventually, some of the sand will

Great Sand Dunes National Park, Colorado.

slide down the slip face in the form of a small avalanche, thus adding a thin layer of sand to it. So, when wind blows, sand is transported from the windward face to the slip face, and the dune itself thus slowly migrates. Oftentimes, a dune will also migrate onto the windward face of another dune, and winds may also change direction over time, leading to a reversal of their form, as well.

In any of these situations, **cross-bedding** can be developed, which is seen as horizontal units of sediment or sedimentary rock that are internally composed of inclined layers.

The presence of large cross-bedded layers in outcrops of sandstone is a good indication that the material was part of a dune at one time, regardless of what the current climate in the area is.

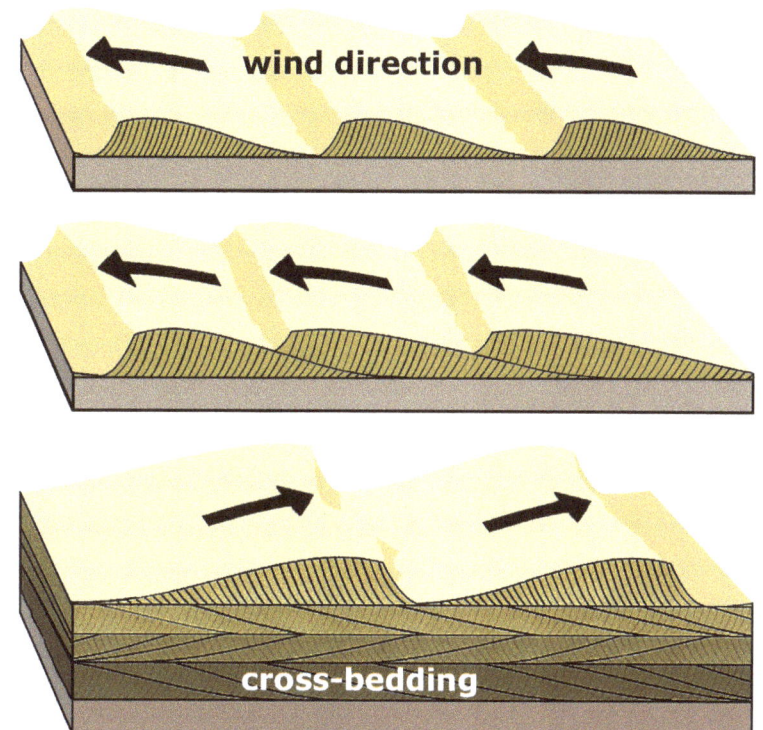

Cross-bedded sand in a dune at Great Sand Dunes National Park, Colorado.

Cross-bedded sandstone in an ancient dune at Valley of Fire State Park, Nevada.

As winds blow across desert areas, they can also create dust or sandstorms, in which large amounts of fine sediment are lifted into suspension and saltate across the ground. These sediments eventually settle out of the air, sometimes great distances away. Thus, winds can move large amounts of relatively fine material out of a desert area.

Winds can modify the landscape in other ways, too. Of course, they can weather and erode material through abrasion, often creating unique geologic features. And, they can lower surfaces and leave them covered with a veneer of coarse sediments called **desert pavement**.

A large dust storm in Iraq.

Pavement can be produced by winds where loose sediments of significantly different sizes are found together, because the wind will tend to remove only the smaller and lighter sediments, while leaving the coarser and heavier sediment behind. This preferential removal of finer sediment causes the surface to drop, and the average grain size of the remaining sediment to increase.

This size-selective erosional process is called **deflation**, which is the removal of loose sediments and the lowering of surfaces by wind.

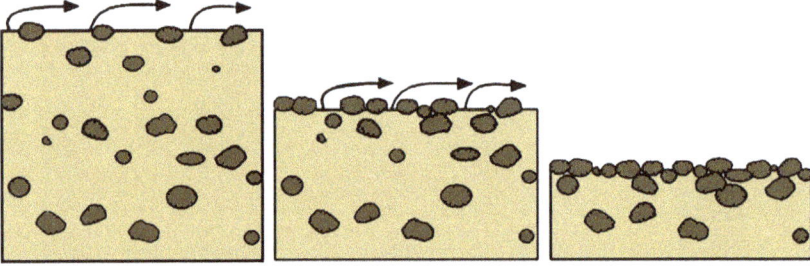
When deflation occurs, surfaces are lowered and become coarser as finer sediments are blown away.

Desert pavement in Death Valley National Park.

Death Valley from space. There's nowhere else like it on Earth.

Unit 3

Section 3.1: The Earth's Structure and Plate Tectonics

The Earth, like the other planets, has a layered structure, and its major parts vary in composition. The outermost layer, the crust, is also comprised of a number of pieces called lithospheric plates, which vary in composition, as well. These pieces move about over time, albeit very slowly, and interact with each other in differing ways, with the interactions of lithospheric plates being known as **plate tectonics**.

The Earth's Structure:

When the Earth initially formed over millions of years, it was a molten mass. It was heated by the repeated impact of objects, the incorporation of decaying radioactive material, and gravitational pressure, but eventually began to cool. As it did, large quantities of gas were released, forming the planet's early atmosphere, while other materials began to separate out due to their differing densities. This process of melting, outgassing, and separation of materials, is called **planetary differentiation**. As differentiation proceeded and the Earth continued to cool, denser materials sank, while less dense materials rose, and the planet became internally layered.

Inner core:

Outer core:

Mantle:

Crust:

Continental crust:

Oceanic crust:

Lithosphere:

Asthenosphere:

Plate Tectonics:

In the early 1900's a scientist, Alfred Wegner, came up with the idea that, at some point in the past, all of the continents existed as a single "super continent" called **Pangea**, which had broken apart over time. The idea was that Pangea had rifted into several large continental blocks, and that these blocks had slowly drifted away from each other until reaching their current locations. Wegner used several independent lines of evidence to support his idea, which led to the development of the **Theory of Continental Drift**.

These four lines of evidence were:

#1:

#2:

#3:

#4:

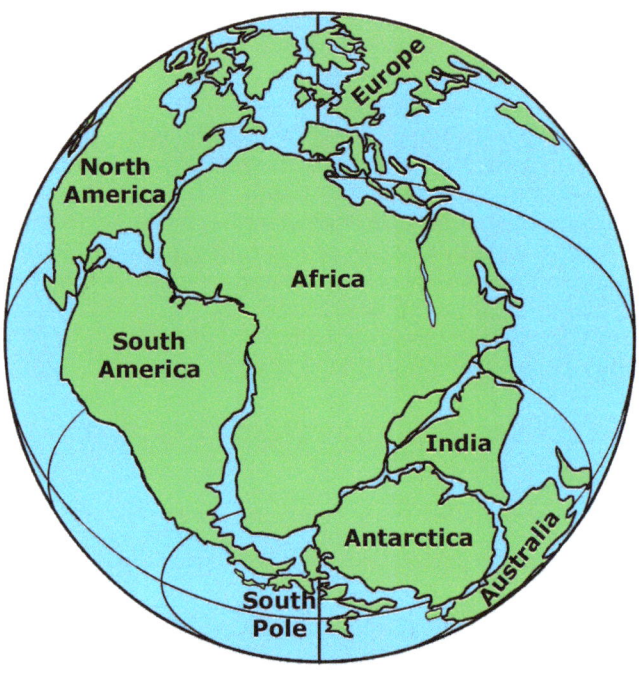

Pangea, as it began to break apart about 200mya.

#1 The overall form of the continents fit together quite well in several places. For example, the continental shelves of South America and Africa match almost perfectly.

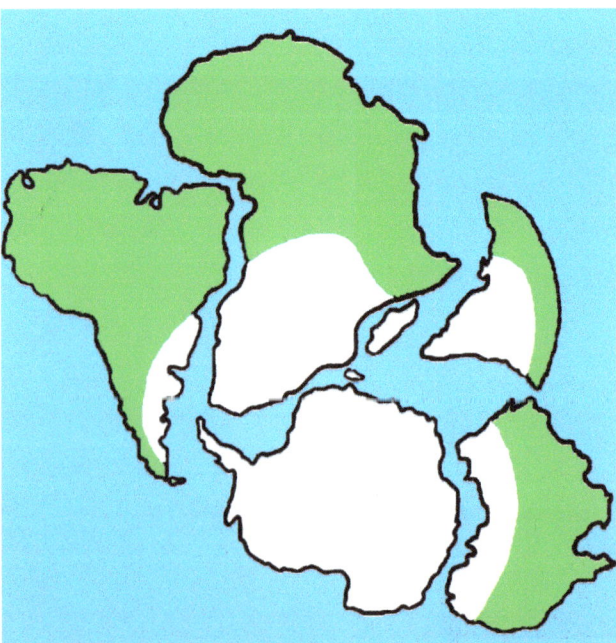

#2 Paleoclimatic evidence in several areas indicates that the Antarctic ice sheet/glaciers spread from Antartica onto other continents,

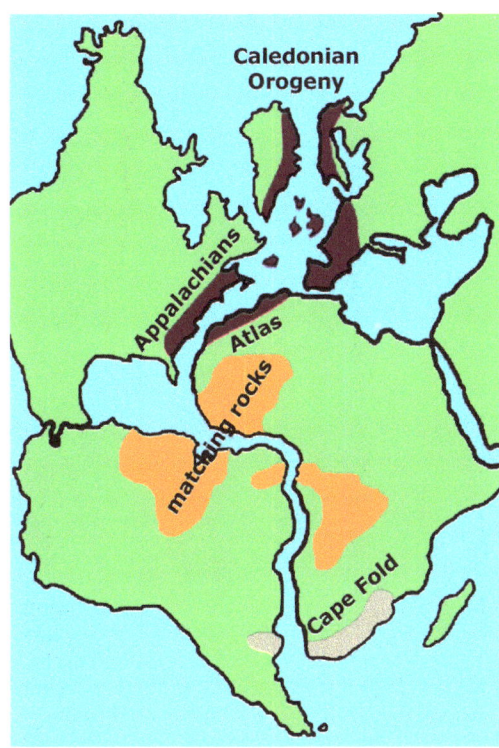

#3 Some mountain ranges and areas with particular types of rocks also match very well across different continents.

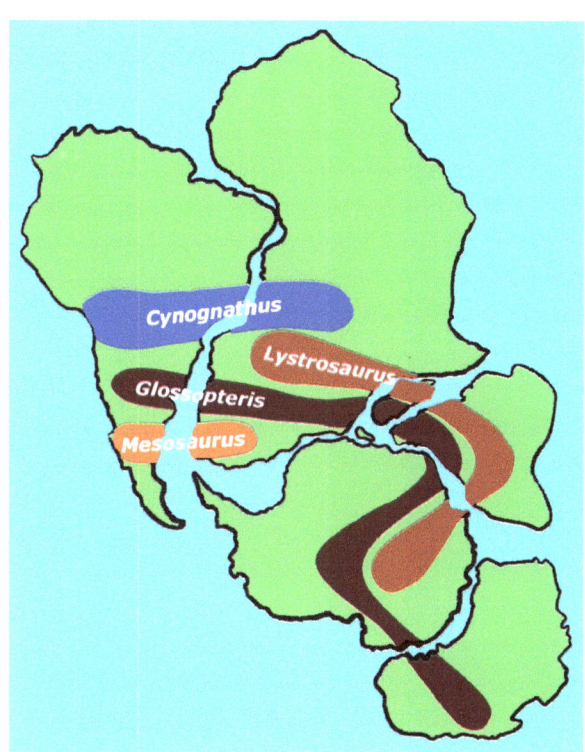

#4 Numerous organisms lived only in certain areas of more than one continent, with their ranges matching very well across them.

Over time, more evidence that supported this idea was found. However, using new technologies, it also became clear that the ocean floors moved, as well. In fact, it was determined that the crust was broken into numerous pieces called **lithospheric plates** that included portions of the crust along with the upper mantle underneath them, which all moved and interacted. These pieces are also called **tectonic plates**, and this additional supporting evidence led to the development of the **Theory of Plate Tectonics**.

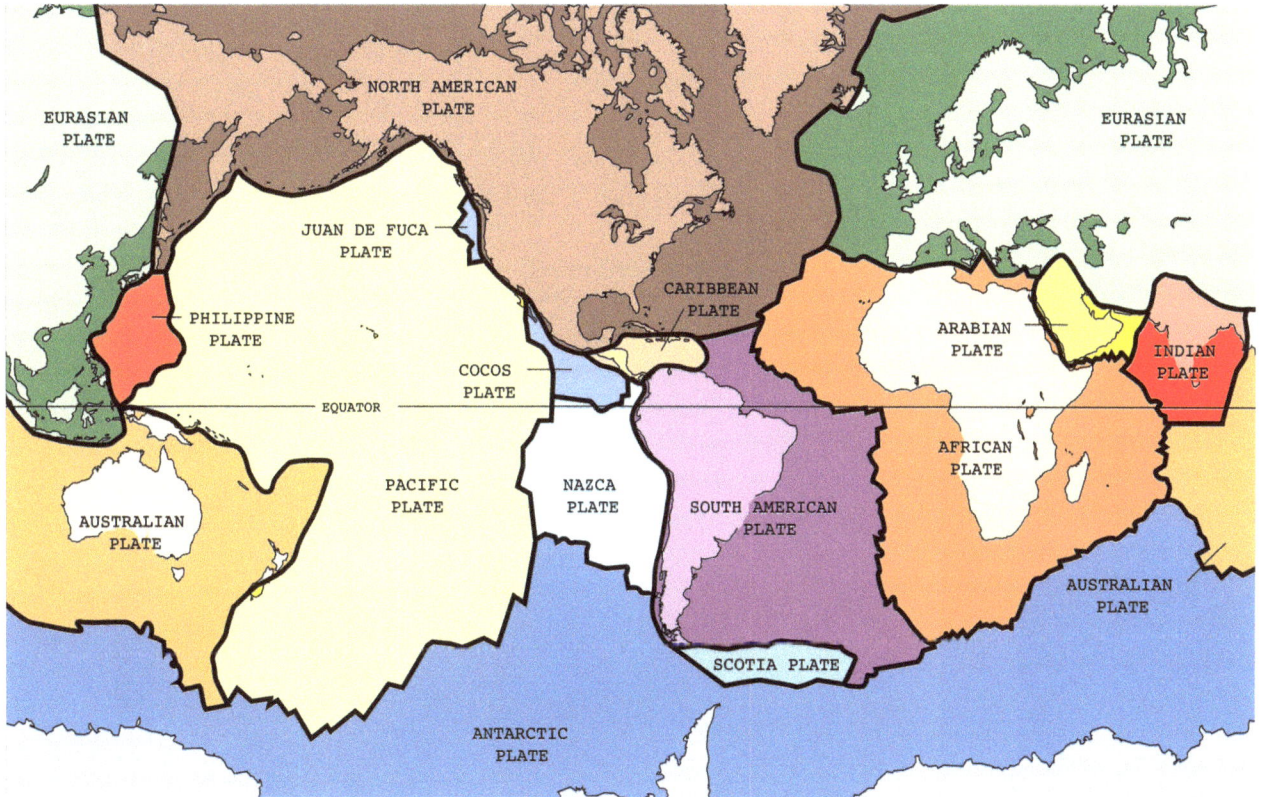

Earth's major tectonic plates, all of which move and interact with those around them.

The four additional lines of evidence are:

#1:

#2:

#3:

#4:

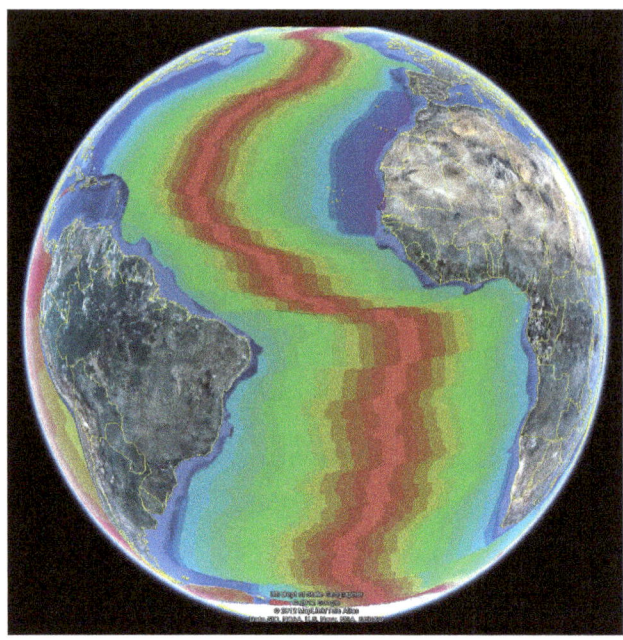

#1 Rather than being the same age everywhere, oceanic crust is new/young near oceanic ridges, and increases in age away from them in both directions.

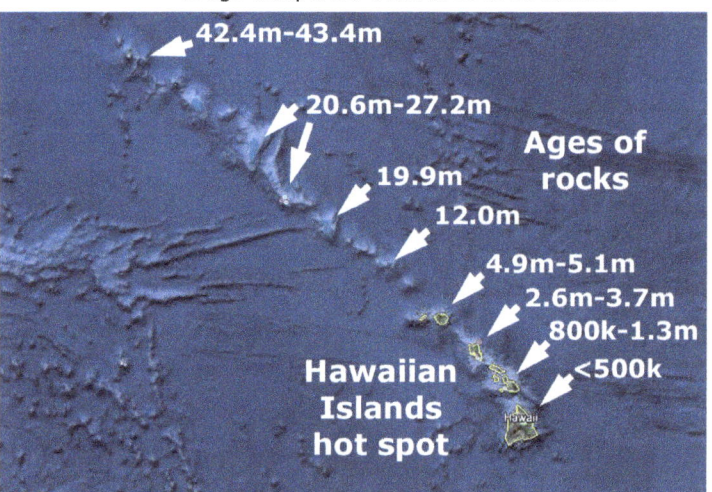

#2 A large body of magma, called a hot spot since it penetrated the crust, produced the Hawaiian Island chain as the Pacific Plate moved northwest over it.

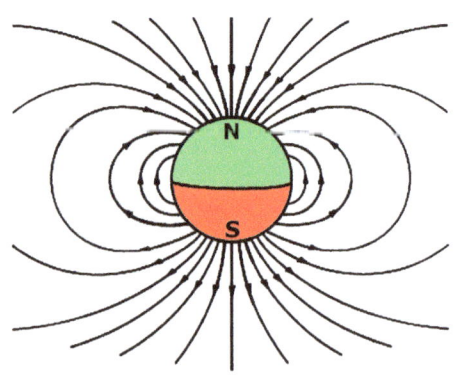

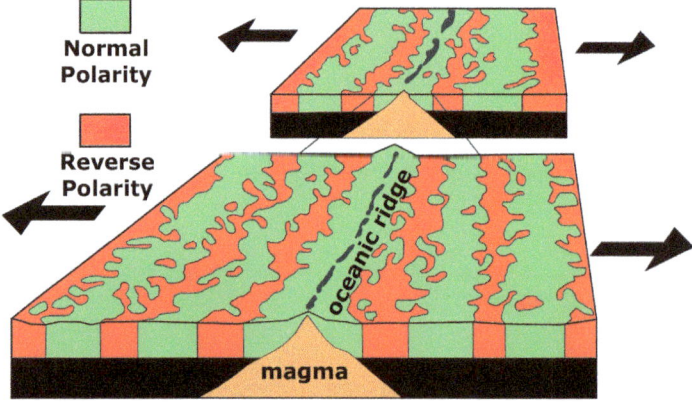

#3 Earth's magnetic field flips, and thus reverses the poles from time to time. This leaves a corresponding magnetic signature in igneous rocks as they form, which creates mirrored bands of rocks with normal and reverse polarity on either side of an oceanic ridge. Magnetic signatures also vary in orientation when rocks are formed as a plate rotates.

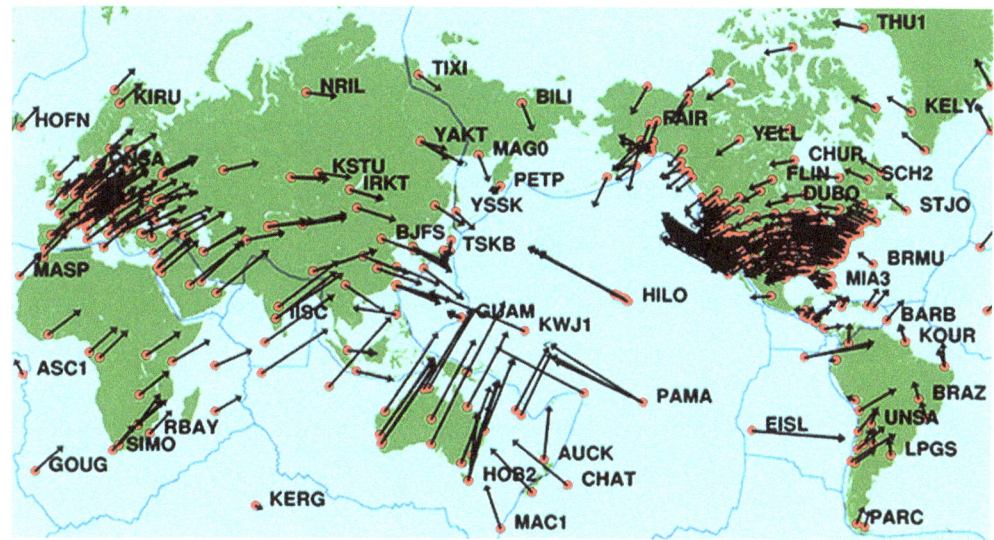

#4 Using technologies like G.P.S. and Very Long Baseline Interferometery, actual plate motion can be measured with great accuracy. The arrows indicate direction of motion, and their length is proportional to rate of motion.

Tectonic Plate Boundaries:

Overall plate motion is driven by convection within the mantle and by the sinking (subduction) of relatively old and dense oceanic crust at some locations. Regardless, plates can interact with each other at three types of plate boundaries. A **divergent plate boundary** is where plates move away from each other and create new crust. A **convergent plate boundary** is where plates move toward each other and collide. And, a **transform plate boundary** is where plates slide past each other laterally.

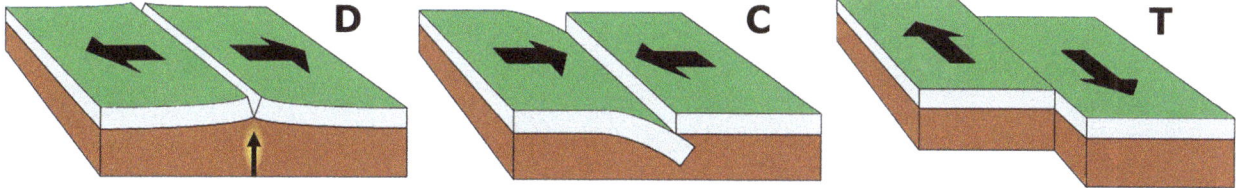

Diagrams of a divergent plate boundary (D), a convergent plate boundary (C), and transform plate boundary (T).

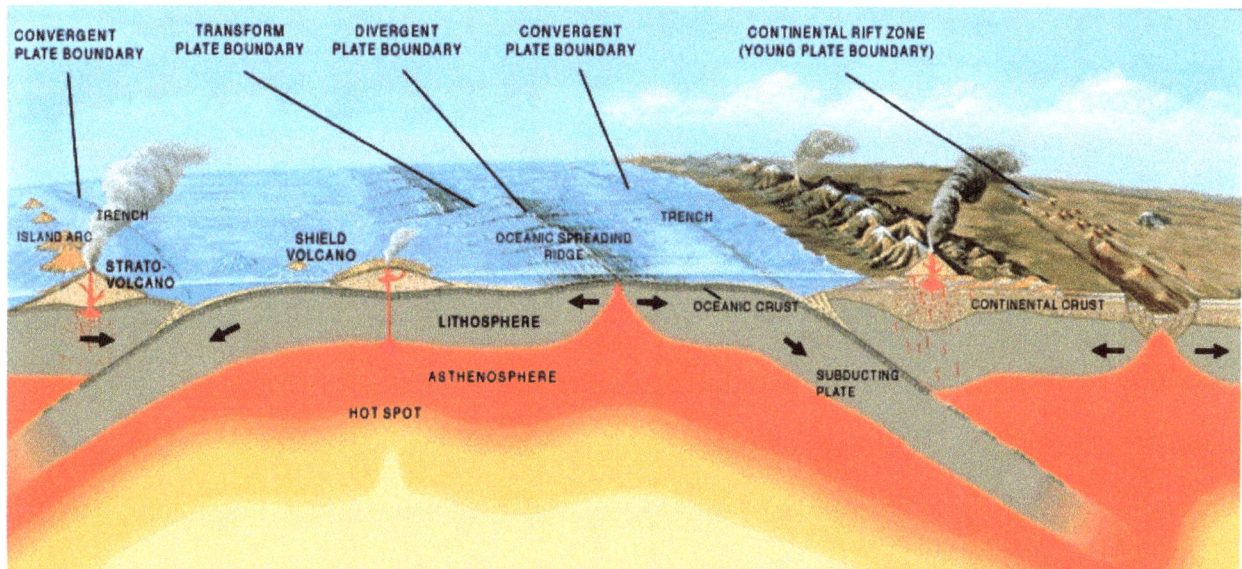

The three types of plate boundaries shown in cross-section, and some associated features.

77

Oceanic divergent plate boundary:

Features:

Example:

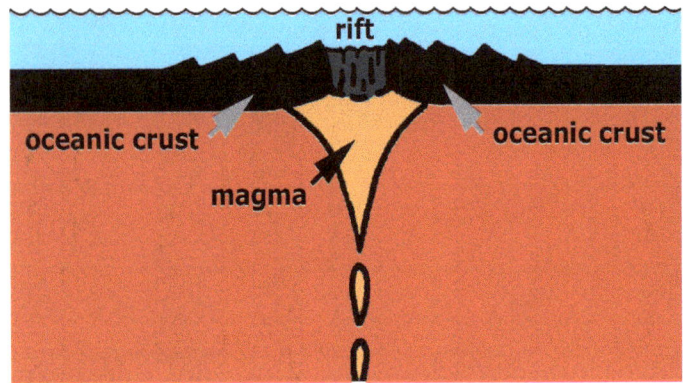 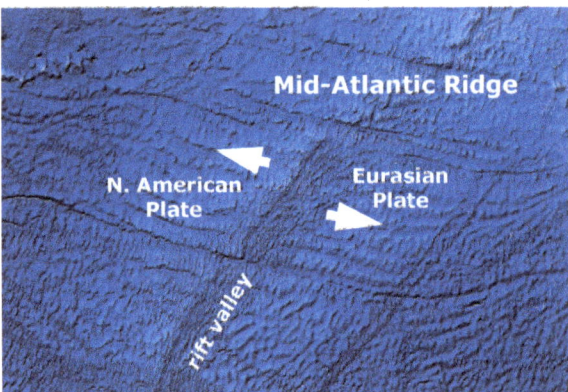

Continental divergent plate boundary:

Features:

Example:

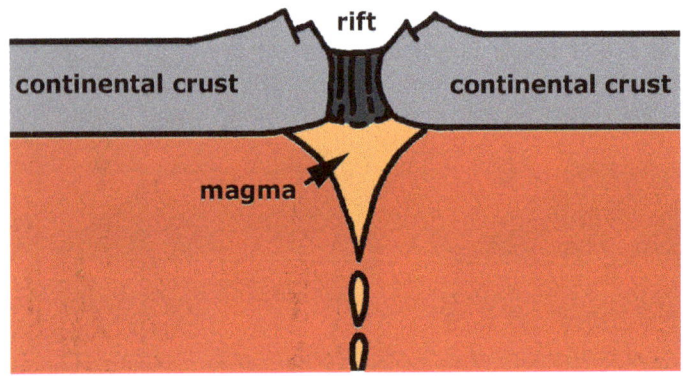 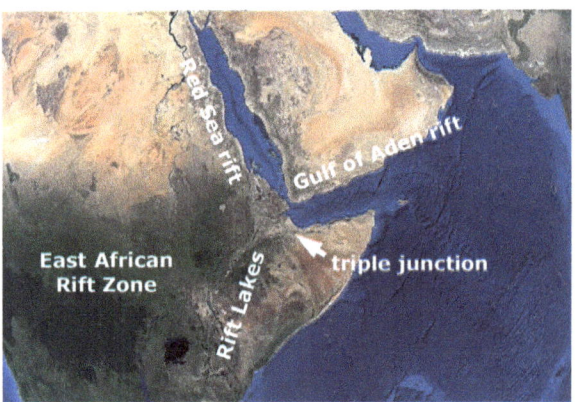

Oceanic-continental convergent plate boundary:

Features:

Examples:

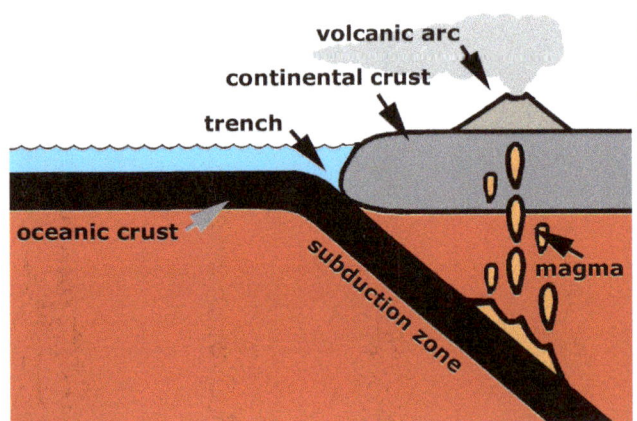 

Oceanic-oceanic convergent plate boundary:

Features:

Examples:

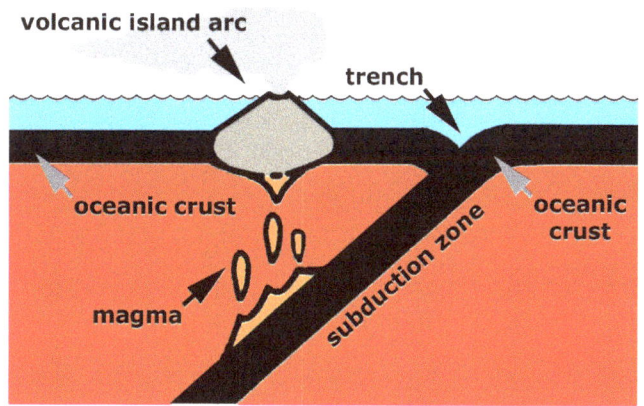
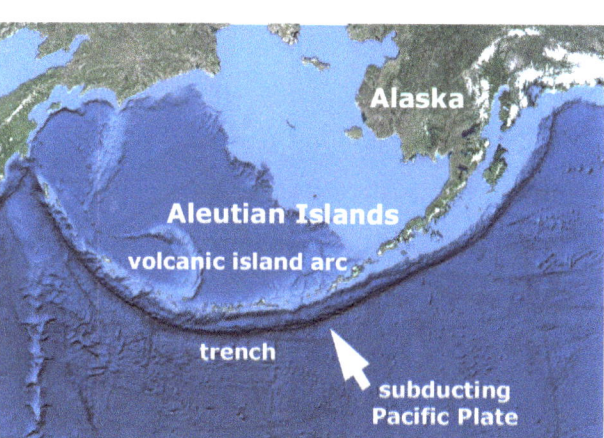

Continental-continental convergent plate boundary:

Feature:

Example:

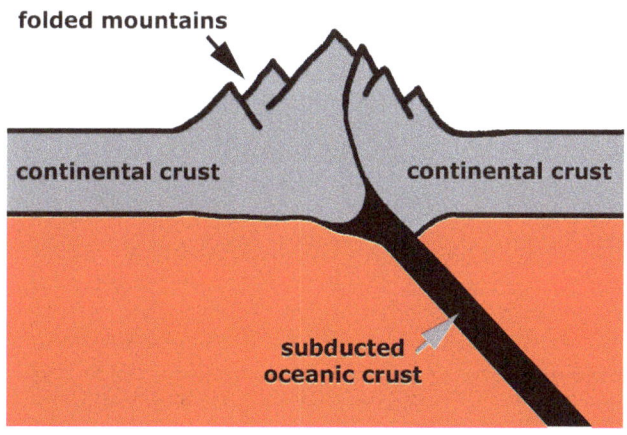
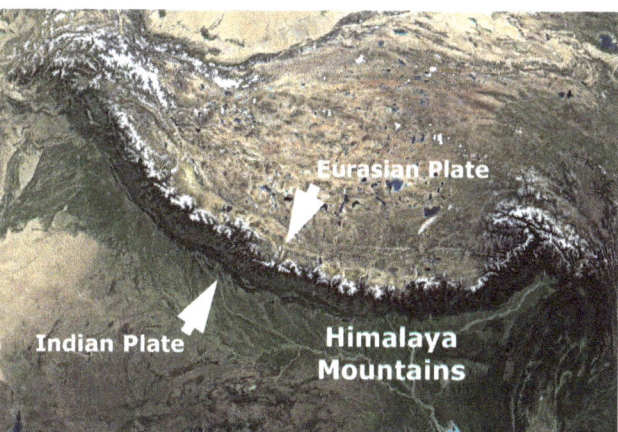

Continental-continental transform plate boundary:

Feature:

Example:

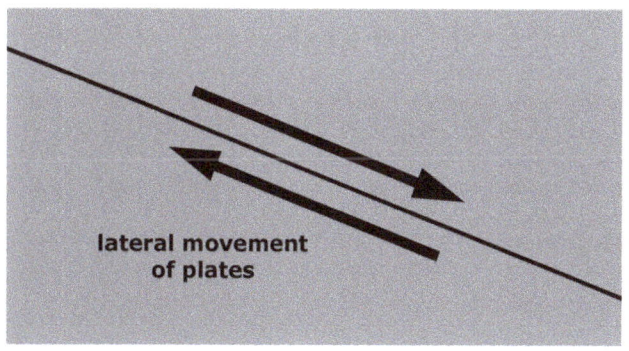

The Scientific Method:

Before moving on, it is important to understand why the Theory of Plate Tectonics is a theory, and why it is inappropriate for some people to claim that it's not well-understood or not well-established because it's "just a theory". So, it is important to understand how science and the scientific method work.

The **scientific method** is the method by which phenomena are investigated, empirical or measurable evidence is generated, and new knowledge is acquired, or previous knowledge is corrected. In other words, it's the way that scientists use evidence to build up our knowledge of the world around us and how it works. There are several steps, although specifically how knowledge is gained can vary depending on circumstances and the nature of the subject being studied. These steps, in general, are:

#1 Ask a question: A scientific study may be broad to narrow in scope, but a detailed question about exactly what a scientist (or group of scientists) wants to find out should be the starting point of any research.

#2 Collect information and make observations: Before doing any sorts of tests, experiments, etc. a scientist should first do some searching and find out anything they can about the subject at hand. They should see if the same, or any similar question, has already been asked, studied, and answered. If so, then they should gather any/all pertinent information. Next, they should make observations of the subject at hand and record what they find, with the goal being the development of an un-tested answer to their question.

#3 Develop a hypothesis: After doing their homework and making and recording observations, a scientist should use what they have learned to produce and un-tested answer to their question, which is called a **hypothesis**. As the hypothesis is developed, it should be made so that it can be tested by some means that can produce empirical or measurable evidence, it should be able to lead to predictions, and it should be falsifiable. For example, a chemist doing research could use their knowledge of chemistry to make a hypothesis that states that if chemicals A, B, C, and D are combined in a specific manner, chemical E should always be produced. This would be a prediction. Likewise, their hypothesis should state that if combining chemicals A, B, C, and D in such a manner produces any chemical other than E, then the hypothesis is incorrect. This would make it falsifiable.

#4 Test the hypothesis: Once a hypothesis is developed, it should be thoroughly tested. This could be done by several means, including physical experiments, computer simulations, etc. In the case of the chemist, it would be time to actually combine chemicals A, B, C, and D, and see if chemical E is produced, or anything else.

#5 Accept, modify, or reject the hypothesis: After thoroughly testing the hypothesis, a scientist should compile the results and determine if the prediction was correct. If so, then the hypothesis has been verified and should be accepted. However, if the results are not what was predicted, then the hypothesis has been falsified, and it should be rejected, or possibly modified. So, if the chemist combined chemicals A, B, C, and D in a specific manner in their lab and only chemical E was produced, then their hypothesis has been verified. But, if anything other than chemical E was produced, then the hypothesis has been falsified. Still, if it was found that, for example, only chemical E was produced as long as only very small amounts of chemicals A, B, C, and D were combined but not when larger amounts were combined, then the hypothesis may be modified to include this information. Of course, this would lead to the creation of a new question regarding why this occurred.

In the case of plate tectonics, a question could be "Does new oceanic crust form at mid-ocean ridges and then move away from them over time?" And, the hypothesis could be "Oceanic crust is produced at mid-ocean ridges and is relatively young there, then moves away from them, with its age increasing with distance from the ridge." Then, to test the hypothesis, samples of oceanic crust could be collected from different areas of the seafloor and dated. If the crust is found to be new/young at the ridge and increasingly older away from the ridge, then the hypothesis has been verified, and should be accepted. Conversely, if the crust is found to be old near the ridge and younger further away, or if the ages of samples are scattered and without pattern, then the hypothesis has been falsified, and should be rejected.

Scientific Theories:

So, how does a verified hypothesis become a scientific theory? To become a scientific theory, the hypothesis must be repeatedly tested, and repeatedly verified, and widely accepted by the scientific community. Thus, a **scientific theory** is defined as a well-tested and widely accepted explanation of a given phenomenon, and is basically the best explanation that scientists have for something, based on all of the evidence they have been able to accumulate.

In reality, a scientific theory is often a collection of repeatedly verified hypotheses, though. For example, scientists developed another question and hypothesis that revolved around plate tectonics and the ages of rocks in volcanic island chains, such as the Hawaiian Islands. If the Pacific Plate was indeed moving and passing over a large body of magma as discussed, then the islands would be new/young right over the magma, and would be progressively older with increasing distance from the magma body. Of course, this is the case, and it was verified when rock samples from different islands were collected and dated. Thus, a different type of evidence was developed and used to build up our understanding of how the Earth works.

All of the basic types of evidence presented for continental drift and plate tectonics come together this way, as each is a repeatedly verified hypothesis that, combined with the others, paints a very clear and widely-accepted picture of how plate tectonics works. So, we have the Theory of Plate Tectonics.

Of course, scientific theories are modified or even throw out from time to time, typically when a new technology arrives, which provides additional or conflicting information. It is rare these days, but earlier in the history of science, it was not uncommon.

The Theory of Continental Drift changing to the Theory of Plate Tectonics is a good example. Initially, it was thought that only the continents moved about, which was verified by several different lines of evidence. So, we had the Theory of Continental Drift. However, scientists didn't really know very much about what was going on in the oceans at the time. Then, with the advent of deep-sea drilling and the ability to collect and date samples of oceanic crust and measure their magnetic signatures, etc. scientists eventually determined that it wasn't just the continents doing the moving. So, the Theory of Continental Drift went away, and the Theory of Plate Tectonics took its place. This didn't mean any of the covered hypotheses that supported continental drift were wrong, rather they were insufficient to create a complete picture of how things worked.

Regardless, the Theory of Plate Tectonics is the best explanation of how plates can form, move, and interact with each other, which is based on a large amount of evidence and several repeatedly verified and widely-accepted hypotheses. So, this is why it is inappropriate to think of it, or any other current scientific theory, as "just a theory".

Scientific Laws:

With that covered, also note that scientific theories and scientific laws are not the same thing. While a scientific theory is defined as a well-tested and widely accepted explanation of a given phenomenon, a **scientific law** is defined as a verbal or mathematical statement that expresses a fundamental principle of science that appears to always be true. In the simplest sense, scientific laws, which are sometimes called scientific principles, state something to which there are no exceptions, but they don't provide a mechanism for, or explanation of, phenomena. An example will make this easier to understand, with Newton's Law of Universal Gravitation versus Einstein's Theory of General Relativity being a good one.

The Law of Universal Gravitation states that all objects with mass are attracted to all other objects with mass (due to gravity), and that the attraction increases as the objects' masses increase, but decreases as the distance between the objects increases. Newton also developed a mathematical formula as part of the law, which can be used to accurately calculate exactly how strongly objects are attracted to each other if their masses and the distance between them is known. However, Newton did not know, nor did he ever discover, *why* objects are attracted to one another. He knew they were, and how strongly they were, but not *why* they were.

Over 200 years later, Einstein explained that there was a curvature of space-time that affected the behavior of objects with mass. This concept is far too complex to explain here, but the point is that Einstein eventually produced the Theory of General Relativity (also known as the Theory of Gravitation), which finally explained *why* objects with mass are attracted to each other. He basically explained why gravity exists, which is what Newton couldn't figure out, and that is the difference between a scientific theory and a scientific law.

Section 3.2: Earthquakes

An **earthquake** (also known as a quake, a tremor, or a tremblor) is a vibration of the Earth caused by a rapid release of energy. They typically occur along tectonic boundaries of some kind, with **the majority being produced at subduction zones and along faults**, while some are produced by volcanic activity. And, while most are too weak for humans to feel, a few each year are strong enough to be destructive.

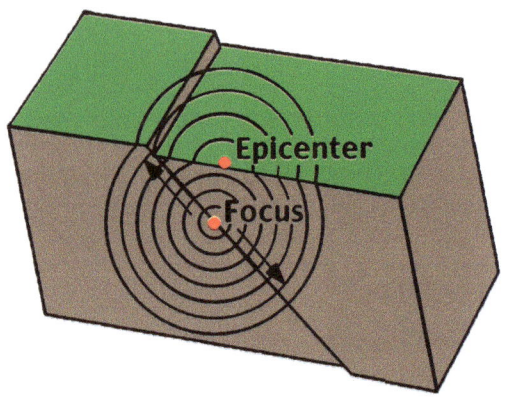

Each earthquake has a **focus** or **hypocenter**, the subsurface area where it actually occurs, and an **epicenter**, which is the location at the surface directly above the focus. And, each produces a few different types of shock waves that radiate from the focus.

Typical earthquakes are the result of rock bodies being subjected to stress, due to movements within the lithosphere, and failing when such stress becomes too great. When failure occurs, the energy stored in the rock body as a result of the stress is released. The easiest way to think about this is to imagine a person trying to bend a thick wooden rod with both hands. Applying pressure to each end of the rod in an attempt to bend it is the application of stress, and when the rod is bent too much, it will suddenly snap and shake the person's hands violently. It will likely shatter to some degree at the breaking point, and a create a loud noise when energy in the form of sound waves is released, as well.

In the case of a subducting plate or fault, two masses of rock are moving relative to each other, and attempting to slide past each other. However, friction and irregularities on their surfaces keep them from doing so. So, they attempt to bend, but end up violently slipping past each other and releasing stored energy. This is called a **stick-slip motion**, as the rock bodies stick to each other until the stress becomes to great for them to do so, at which point slippage (movement) occurs, producing an earthquake.

When slippage occurs, the shock waves produced are properly called **seismic waves**, and these waves come in two basic forms. Some of these waves travel through the Earth, while others spread out across the Earth's surface. So, those that travel through the Earth are called **body waves**, and those that travel across the surface are called **surface waves**. Body waves also come in two forms, which are **primary waves** (P-waves) and **secondary waves** (S-waves).

Primary waves:

Secondary waves:

Because **primary waves travel through materials about 1.7 times faster than secondary waves**, as both types of body waves move away from their source, the primary waves will be the first to arrive at a distant location, with the secondary waves arriving at a later time.

All seismic waves can be detected by an instrument called a **seismograph**, which can produce a paper or digital record of the waves and their intensity as a **seismogram**. So, the difference in arrival time of different types of waves can be seen and measured.

This allows geologists (specifically seismologists) to determine the distance to an earthquake's epicenter by measuring the amount of time that passes between the arrival of the first primary waves and the

A seismograph, producing a paper seismogram.

first secondary waves at a given location, and using these times on a **Travel-Time Graph**. However, using the graph can only determine the distance to an earthquake's epicenter and not the direction. So, data from at least three seismographs must be available to plot an epicenter's actual location. Using data from three seismographic stations, a circle of the correct size (which is found when using a Travel-Time Graph) is drawn around the location of each seismographic station, showing how far away the earthquake occurred from each. Then, the epicenter is found wherever the three circles overlap, with this process being called triangulation.

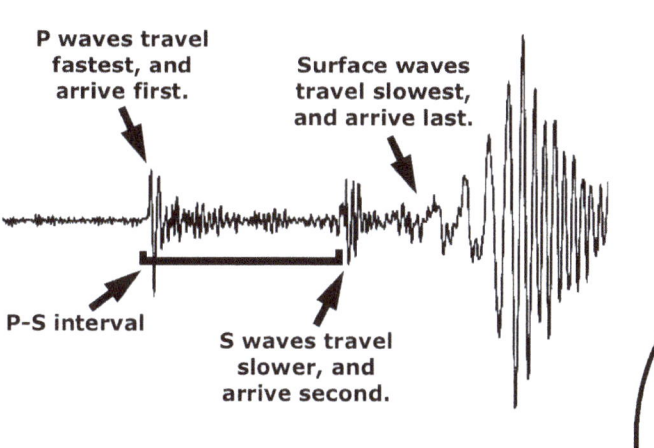

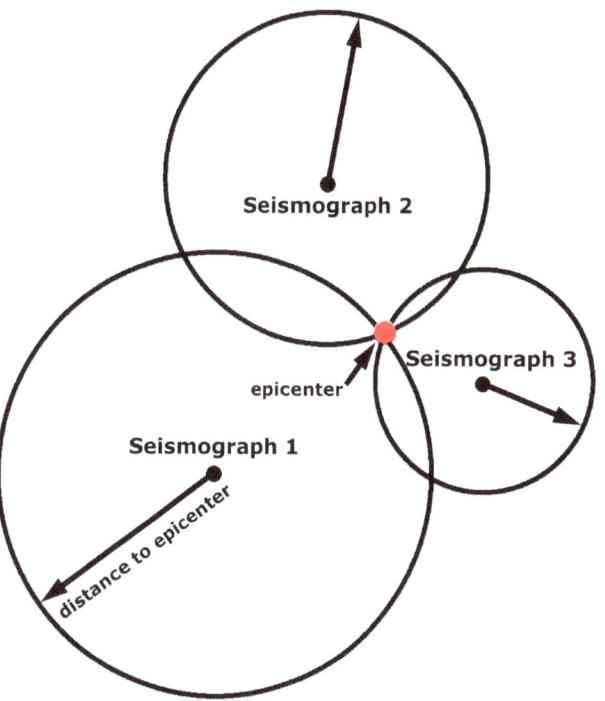

Above, a seismogram showing the arrival of primary waves, secondary waves, and surface waves.

Right, three circles showing how triangulation can be used to determine the location of an epicenter.

Below, a Travel-Time Graph, used by seismologists to determine the distance to an earthquake's epicenter.

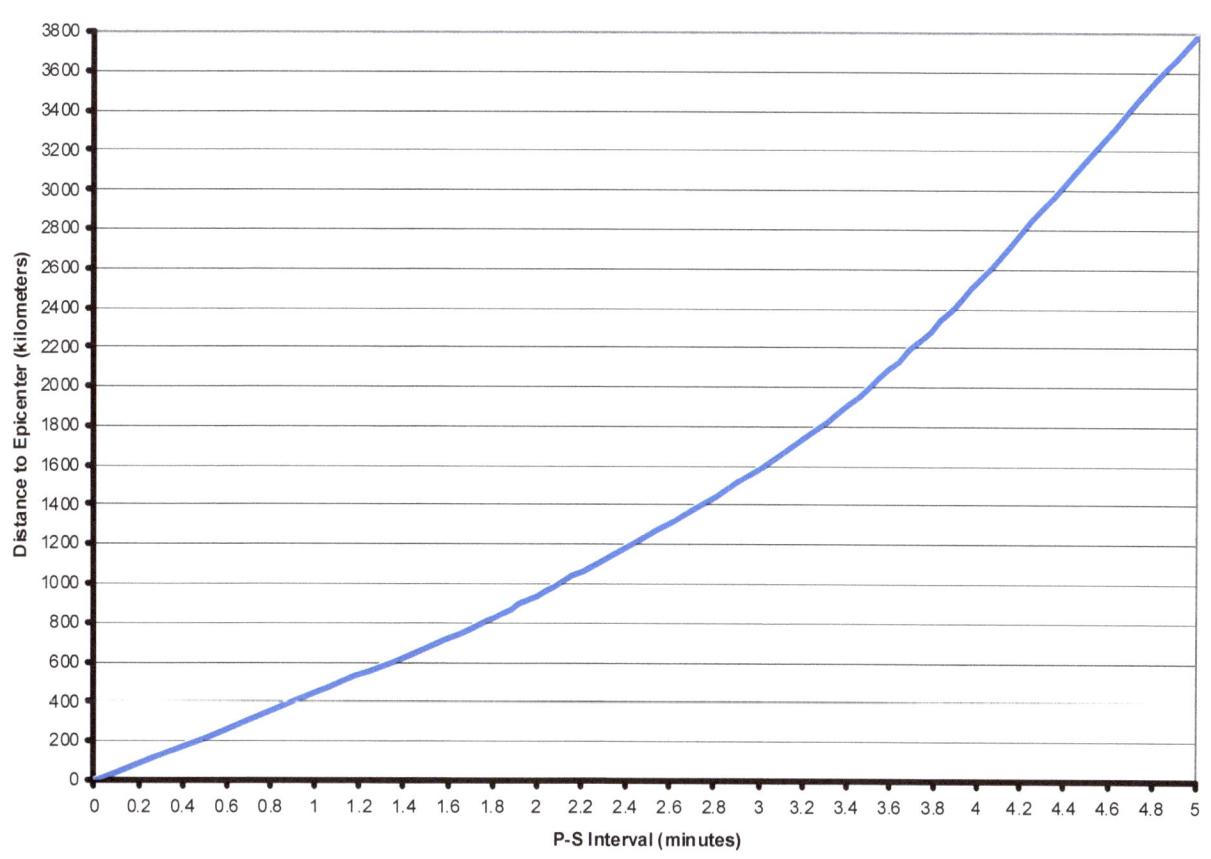

83

In addition to location, an earthquake's strength can also be determined using seismographic data. The strength is properly called an earthquake's **magnitude**, and there are several scales upon which earthquake magnitudes are ranked. However, the **Richter Scale** is the best-known, while the **Moment-Magnitude Scale** is the most commonly used by seismologists.

Again, both measure magnitude, but they do so in different ways, with the Moment-Magnitude Scale being the more accurate of the two. So, while **the most powerful earthquake ever recorded on the Richter Scale was ranked at 8.9, it was ranked at 9.5 on the M.M.S.**

Rankings on either scale can be deceptive though, as an increase of 1 on either equates to a **ten-fold increase in ground shake**. For example, if one earthquake registers as a 4 and another earthquake registers as a 5, this means that the second earthquake shook the ground 10 times more than the first.

With this in mind, it should make sense that a magnitude 1 to 2 earthquake on either scale can't even be felt by humans, while a 4 to 4.9 can be felt by most people in the area, and a 5 to 5.9 produces damaging shocks. That's a significant jump from being felt by most to damaging shocks when going up just one number from a 4 to a 5. But again, that's because a 5 shakes the ground 10 times harder than a 4. Knowing this makes it easy to understand why an earthquake with a magnitude greater than 8 can be catastrophic, as it shakes the ground at least 1,000 times harder than a 5, which can cause minor damage.

In addition to ranking earthquakes by magnitude, they are sometimes ranked by **intensity**, which is their effect on people, human structures, and the natural environment, as well. The **Modified Mercalli Scale** is used in the U.S., and is comprised of increasing levels of intensity from imperceptible shaking to catastrophic destruction based on various effects experienced near the epicenter. In general, the lower end of the scale reflects how an earthquake was felt by people, but the higher end reflects the types of damage done.

Location of Occurrence:

As mentioned, most earthquakes are produced along plate tectonic boundaries, especially where subduction takes place, and along faults. So, **most occur along the Circum-Pacific Belt**, which is commonly called the **"Ring of Fire"**. This is the belt of volcanic arcs and volcanic island arcs that runs around much of the Pacific Ocean basin, where several plates are being subducted.

Magnitude	Ave. per Year
> 8	1
7 - 7.9	15
6 - 6.9	135
5 - 5.9	1320
4 - 4.9	13,000
3 - 3.9	130,000
2 - 2.9	1,300,000

The average number of earthquakes, globally, per year, for given magnitudes.

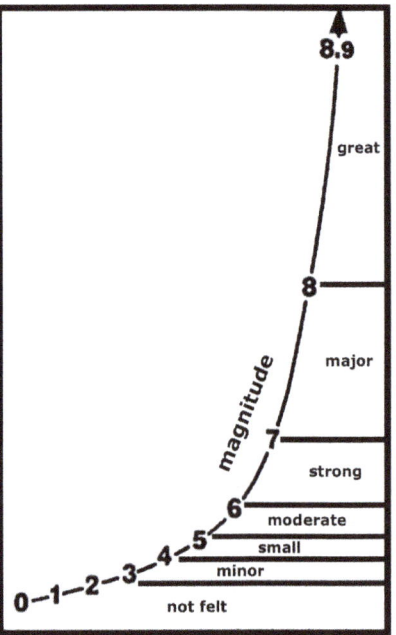

The severity of earthquakes increases rapidly on the Richter and M.M.S. scales.

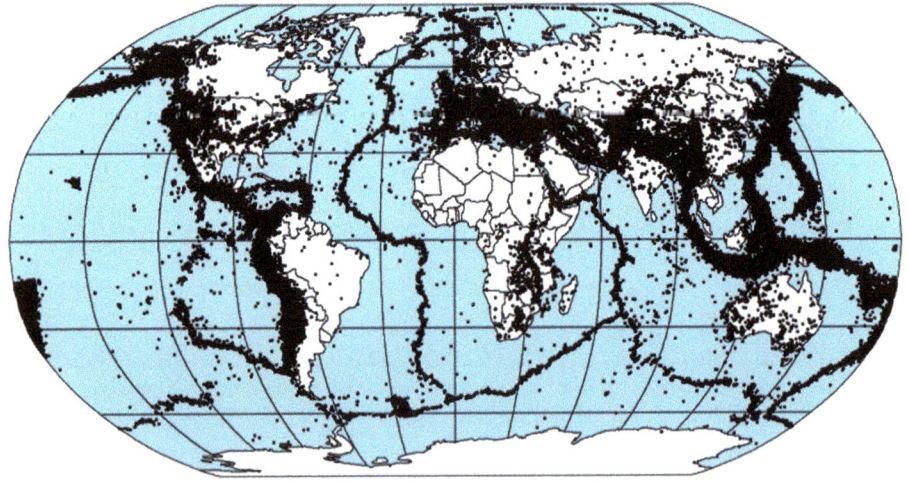

Thirty-five years of earthquakes are shown. Notice how many are clearly associated with tectonic plate boundaries.

However, this is obviously not the only place they are produced, as damaging, sometimes catastrophic, earthquakes have occurred all over the world.

Examples:

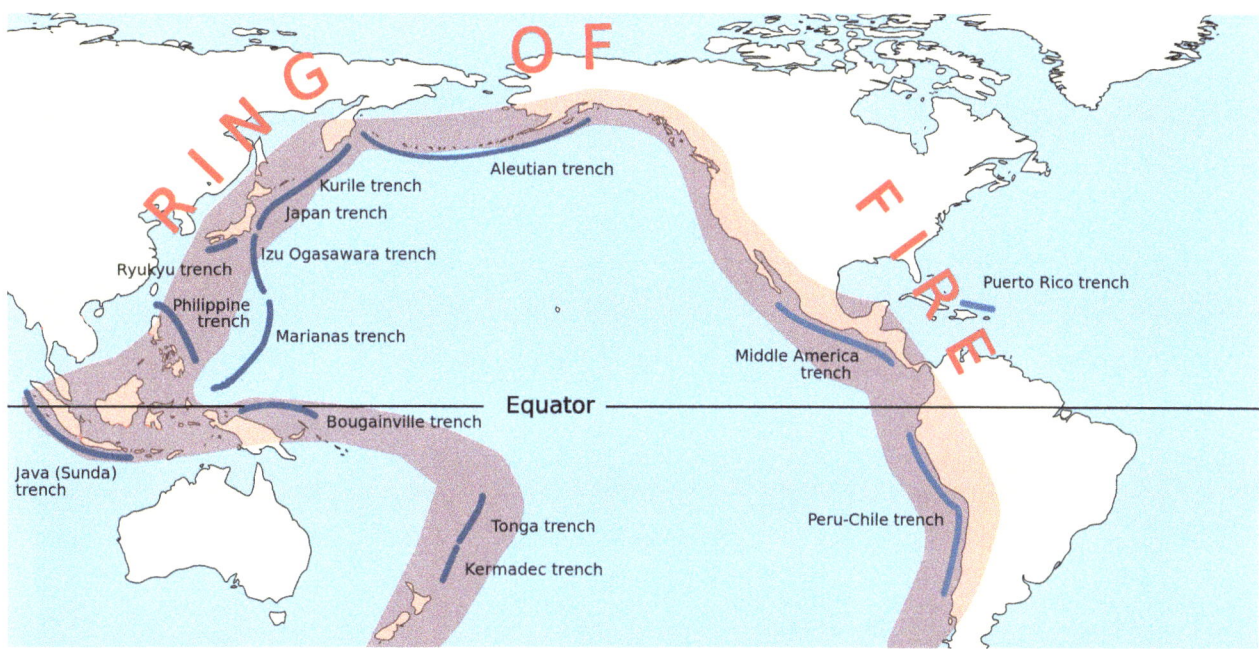

The Circum-Pacific Belt, better known as the "Ring of Fire", and the trenches associated with it.

In the U.S. several earthquakes with magnitudes greater than 8.0 have occurred in Alaska, the southern/southwestern area of which is part of the Circum-Pacific Belt. But, **the most destructive/deadly ones in the lower 48 states have occurred in California, along the San Andreas Fault**, where over a half-dozen earthquakes have had estimated magnitudes greater than 7.0. This fault, which is a transform plate boundary and lacks volcanoes, is still part of the Circum-Pacific Belt. Still, several large earthquakes have also occurred in other areas, including the **New Madrid Fault, Missouri**, and in Charleston, South Carolina. In fact, three of the highest-ranked earthquakes outside Alaska and California were on the New Madrid Fault, having estimated magnitudes of 7.5 to 7.7, while Charleston's was a 7.3.

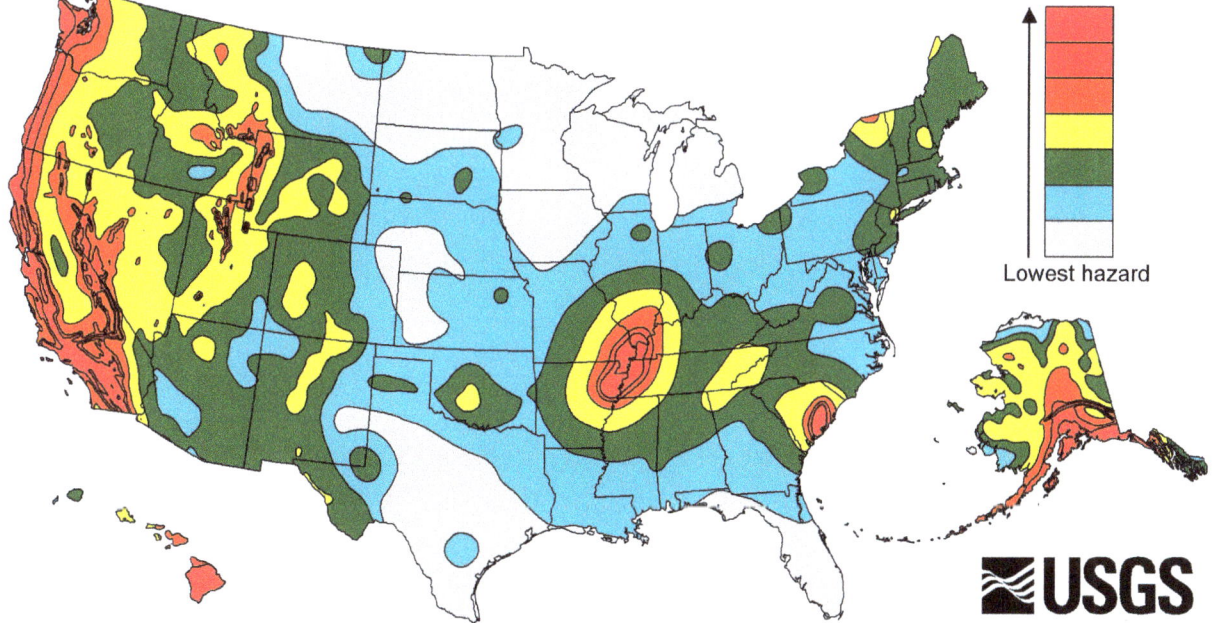

An earthquake hazard/risk map. California and the rest of the West Coast, the New Madrid area, and Charleston are relatively high-risk areas, along with the Yellowstone National Park, Wyoming area.

85

Earthquake Hazards:

#1 Vibrations:

#2 Fires:

#3 Liquefaction:

#4 Tsunamis (seismic sea waves):

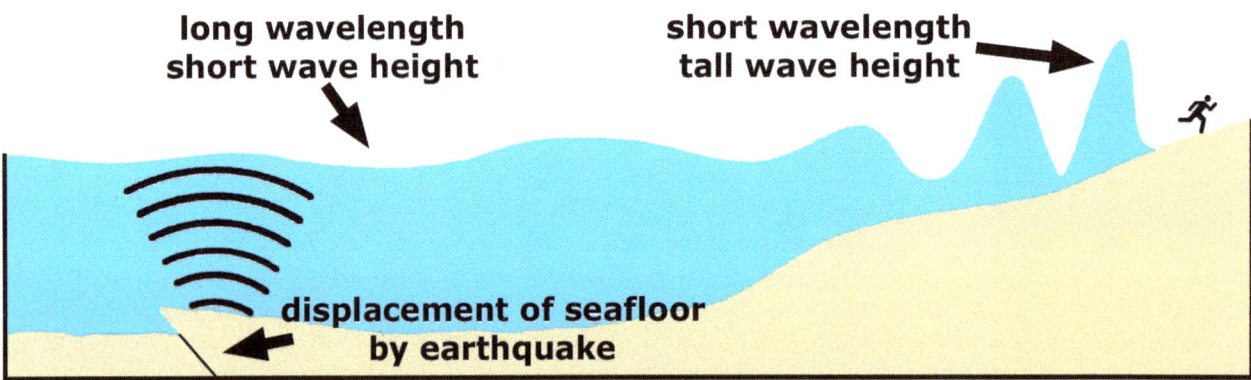

In open/deep waters, tsunamis have an exceptionally long wavelength and a very short wave height. However, upon reaching shallow waters, the wavelength rapidly decreases and the wave height increases dramatically. This is why they often surprise and kill people in low-lying areas, as there is oftentimes little warning, or time to evacuate.

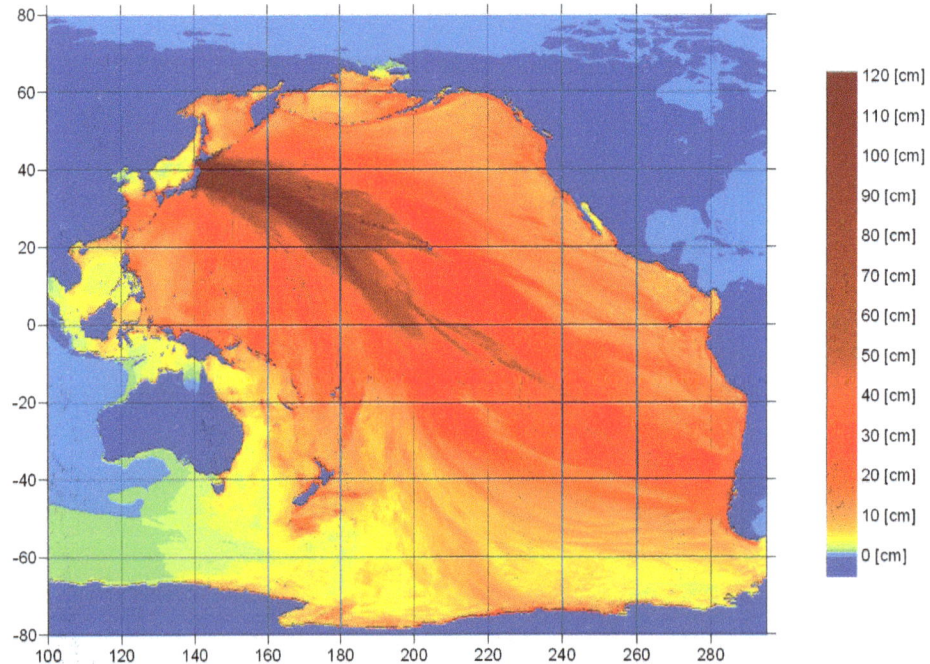

The Honshu, Japan earthquake of 2011 was a 9.0 on the M.M.S. and produced one of the most destructive tsunamis known. Surprisingly, in the deep waters off the east coast, it still had a maximum wave height of only 1.2 meters.

Section 3.3: Volcanoes and Plutons

A **volcano** is a mountain formed by outpourings of lava and/or ejections of solid material, while a **pluton** is any significant structure formed by the subsurface cooling of magma. So, while both are comprised of igneous rock, volcanoes are extrusive features, and plutons are intrusive features. There are a few basic types of each, and all will be covered.

Volcanoes:

There are a few basic types of volcanoes, which are produced by different types of volcanic eruptions, and the differences in their size and form are largely determined by differences in the magma that produces them. The temperature, composition, and gas content of magma can be highly variable. So, all volcanoes are not the same, and they vary in several ways.

Mt. Agung, a stratovolcano in Indonesia.

The temperature and composition of magma affect its **viscosity**, which is its resistance to flow, or thickness. Of course, all magma is very hot, but some is hotter and some is cooler, as it typically ranges from 1,300° to 2,400°F. Magma can also be granitic, andesitic (intermediate), or basaltic in composition. This means that some magma has a relatively high silica content, while some has a much lower silica content. Both of these, temperature and silica content, significantly affect viscosity.

Relatively hot magma:

Relatively cool magma:

Magma high in silica:

Magma low in silica:

Relatively hot and low silica basaltic lava:

Relatively cool and high silica rhyolitic lava:

Basaltic magma is typically hotter and lower in silica than granitic magma, so it usually has a relatively **low viscosity** and commonly forms **lava flows**, which are flowing masses of molten rock, when it reaches the surface. There are two basic types of basaltic flows seen on land, which are produced primarily by differences in temperature and the rate of discharge of lava, as well as the slope of the land it flows over. One type can turn into another if these factors change, as well. In addition, lava may be erupted underwater at times, which produces a third type of basaltic lava flow, which is called pillow lava.

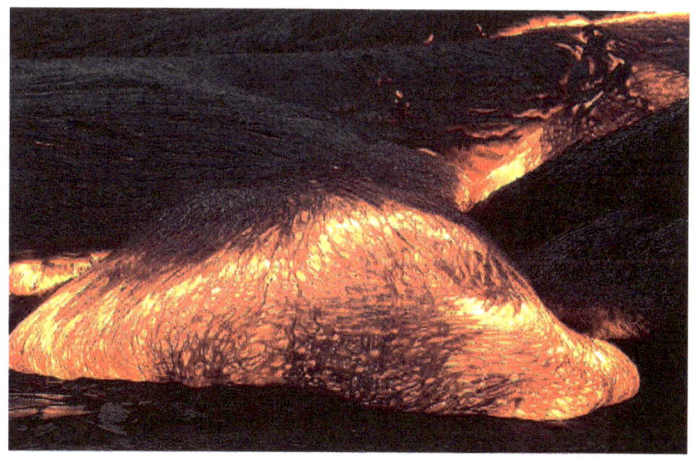

Pahoehoe lava flow:

Aa lava flow:

Pillow lava flow:

Basaltic lava flows may also produce **lava tubes** at times, which are tubes produced when the upper surface of a flow cools and forms an insulating crust of rock, allowing fluid lava to continue to flow beneath it. This fluid lava may keep the tube filled for as long as it is still being erupted, but may then run down slope and out of the tube once an eruption stops. This can leave the tube empty, and looking much like a cavern formed in limestone.

Small and large lava tubes in Craters of the Moon State Park, ID (L) and Hawaii Volcanoes National Park, HI (R).

Rhyolitic lava is typically much cooler than basaltic lava, and is also much higher in silica. So, it doesn't form lava flows the way basaltic lava can. Instead, due to its **high viscosity**, it typically forms **lava domes**, which are roughly circular and mound-shaped hills produced by the slow eruption of rhyolitic lava.

However, on occasion, **coulees** may be produced by rhyolitic lava, or **block lava flows** by andesitic lava, if these lavas are erupted on a slope. These are similar to aa flows, but are covered with large chunks of angular lava. They move very slowly, and do not travel far, though.

A coulee in the Mono-Inyo Craters chain, California.

A rhyolitic lava dome within the Mono-Inyo Craters chain.

Again, magma also contains dissolved gases, which vary in quantity and type. While the complete list is long, **the most abundant gas is consistently water vapor**, which typically makes up over 60% of the gas emitted by volcanoes. **Carbon dioxide is consistently the second most abundant**, typically making up 10 to 40% of emitted gas. Still, significant quantities of harmful gases such as **sulfur dioxide**, may also be emitted. This gas and a few others can have negative effects on human health, can form acid rain, and can damage vegetation/crops. For that matter, carbon dioxide can also adversely affect human health, as it can be deadly when at sufficient concentrations. This normally harmless gas can accumulate in low-lying areas, making it difficult to breath, and can cause dizziness, unconsciousness, or in worst cases, **death by suffocation**. Carbon dioxide is also a greenhouse gas, and thus affects the Earth's climate, although humans produce more than 100 times as much CO_2 every year as all the world's volcanoes combined.

Regardless, when molten rock has a **low viscosity**, these gases typically produce large quantities of bubbles in it at/near the surface. **Much gas escapes** as these bubbles form and then pop, but many are not able to pop if lava cools too quickly. As covered earlier, this leads to the development of the igneous rocks scoria and pumice, which have a vesicular texture.

As gas is released by magma, it can also forcefully expel lava from the ground/a volcano, essentially the same way that CO_2 can forcefully expel champagne from a bottle when the cork is removed and pressure on the contents is quickly reduced. In the case of volcanoes, this may produce spectacular **lava fountains**, which are jet-like vertical eruptions of lava.

On the other hand, when magma has a **high viscosity**, it can effectively clog a volcano and **prevent gas from escaping**. This can lead to a buildup of pressure and may result in a violent **explosive eruption**, in which a significant portion of a volcano may be blasted away, sending huge amounts of material into the sky.

Volcanoes can also produce four types of **pyroclastics**, which are solid materials that travel through the air. These may be solid rocks that are thrown into the air, or may originate as globs of high-flying lava that harden as they travel through the air.

A lava fountain.

#1 Volcanic ash/dust:

#2 Lapilli/cinders:

#3 Blocks:

#4 Volcanic bombs:

A volcano is produced when a body of magma finds its way to the surface and is erupted as lava and/or pyroclastic material. So, no matter their size when fully-developed, they all start as an opening in the ground. The opening that material is erupted from is called a **vent**, and its connection to the magma chamber is called a **pipe**. As lava and/or pyroclastics accumulate around the vent, the volcano increases in size, creating a hill or mountain called a **cone**, and also develops a steep-walled depression at its peak, which is the **crater**. Lastly, the sides of a volcano are called its **flanks**.

In addition to these basic parts, smaller vents may open on the flanks of a volcano, which can produce flank eruptions and **parasitic cones**. These are essentially smaller volcanoes that are fed by the same magma chamber. Linear cracks may also open up on the flanks and in the surrounding area, which can produce elongated **fissure eruptions**. New volcanic cones may develop along these, too. And, openings are sometimes formed from which gases are emitted, but not lava. These are called **fumaroles**.

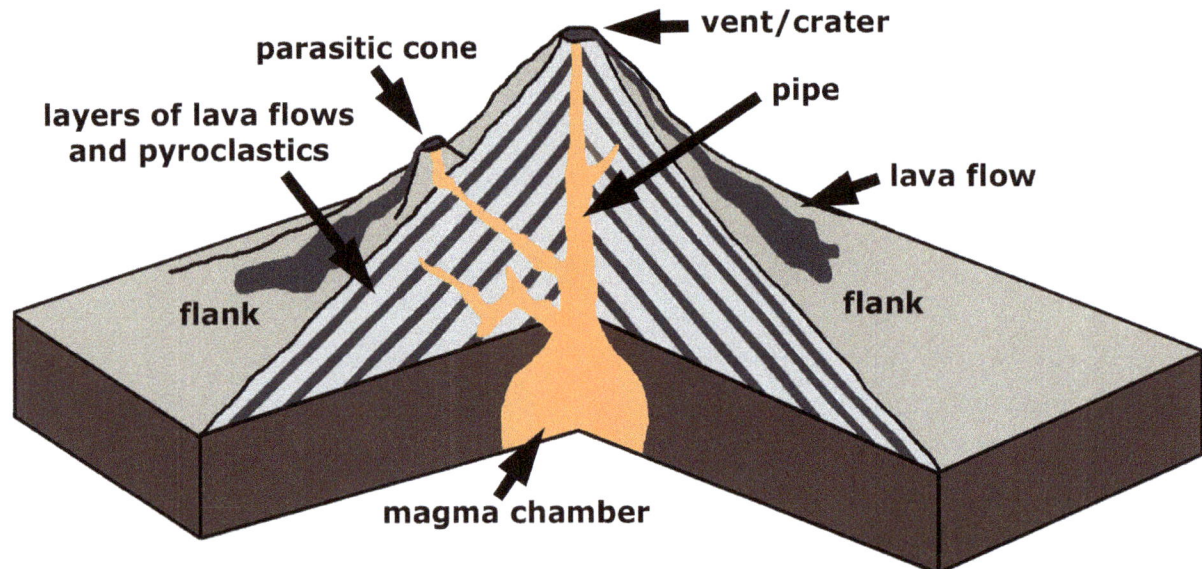

There are three basic types of volcanoes, which are shield volcanoes, stratovolcanoes, and cinder cones. These are produced by the eruption of different materials, and are comprised primarily of lava flows, pyroclastics, or a combination of both. They also vary dramatically in size once fully-developed.

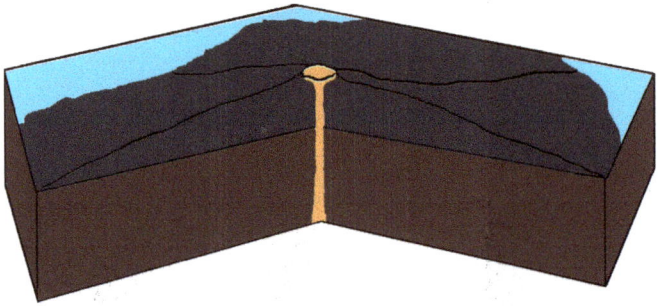

Relative sizes of the three basic types of volcanoes when fully developed.

#1 Shield volcano:

Example:

The shield volcano, Mauna Loa, is the largest mountain on Earth at 30,085' tall. About 14,000' are above water.

Large lava flows from Mauna Loa, which often flow many miles, helped build Hawaii Island in a half-million years.

The shield volcano, Skjaldbreiður, in Iceland with parasitic cones in the foreground.

#2 Cinder cone:

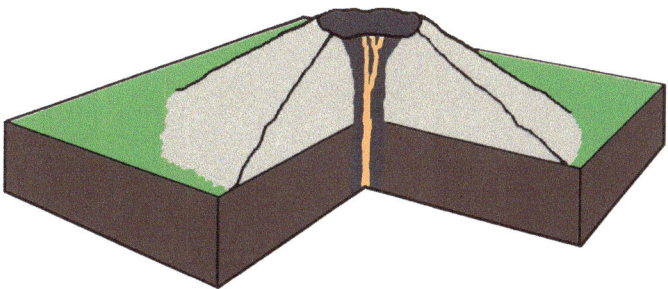

Examples:

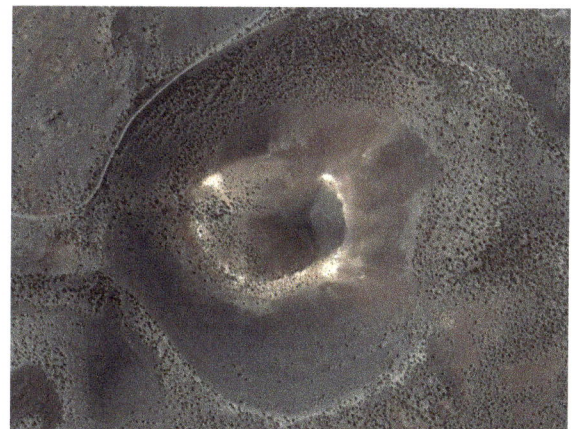

The cinder cone, Sunset Crater, in the Sunset Crater Volcano National Monument, Arizona. It's about 1,000' tall.

The cinder cone, SP Crater (Shit Pot Crater), in the San Francisco volcanic field, Arizona. It's about 820' tall, and is unusual because it created a large lava flow that emerged from the base of the volcano (and for its name).

#3 Stratovolcano or Composite Volcano:

Examples:

The stratovolcano Mt. Fuji, near Tokyo, Japan.

Mt. Fuji is famous simply for its beauty and size. It's 12,388' tall, and the highest peak in Japan.

The stratovolcano, Mt. St. Helens, in Washington State is famous for exploding in 1980.

The explosion reduced St. Helens' height from 9,677' to 8,363' and caused widespread destruction for miles, including snapping these trees at a distance of over 5 miles.

The stratovolcano, Mt. Vesuvius, at only 4,203' tall, is famous for erupting and burying the Roman city of Pompeii in 79 A.D. This killed over 1,000 people, three of which are below.

The last volcano-related feature to cover here is a **caldera**, a large cauldron-like depression formed by the collapse of a volcano into the magma chamber underneath it, typically after a large eruption. When a large eruption occurs, the magma chamber may become partially emptied, and the mass of a volcanic mountain overhead may be too great to be supported. So, the volcano breaks apart and collapses. Afterwards, the remaining magma sometimes produces new volcanic cones in the caldera, too.

The formation of a caldera, as a volcano erupts and then collapses.

Example:

Crater Lake in Oregon, formed when Mt. Mazama collapsed and created a caldera 6 miles across. Wizard Island, a cinder cone, formed afterwards, and over time the caldera filled with water to form the lake.

Volcanic Hazards:

#1 Fires:

#2 Explosions and flying debris:

A volcanic bomb shelter in Japan.

#3 Gases:

#4 Ash:

#5 Pyroclastic flows:

Tens of feet of ash can settle on the surrounding landscape.

#6 Lahars:

A large pyroclastic flow.

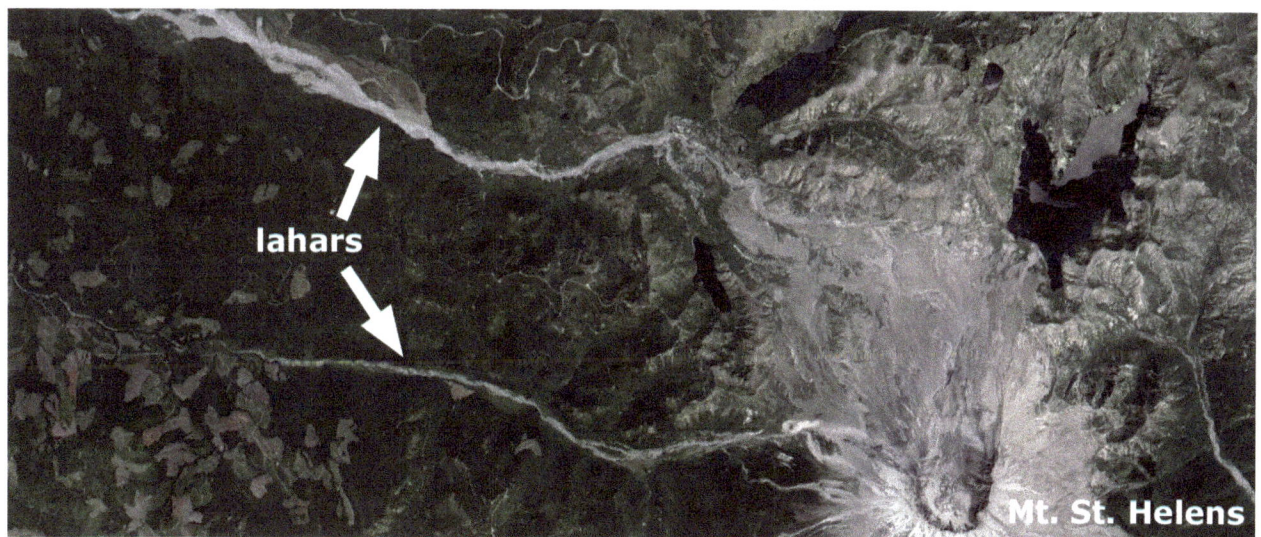

Plutons:

Again, **plutons** are any significant structures formed by the subsurface cooling of magma, and are thus intrusive igneous features. As a body of magma makes its way upwards into surrounding rock, it will typically heave the rocks in the area and fracture them to some degree. Then, magma works its way into any cracks or other areas of weakness and cools, forming intrusive igneous rock.

Some of these plutonic features may cut through/across layers of pre-existing rocks, and some may form between them. So, plutons can be either **discordant** features that cut through/across layers of rock, or **concordant** features that form between layers. Plutons are also either **tabular** and flat in overall form, or **massive** and not flat in overall form.

Dike:

Sill:

Laccolith:

Batholith:

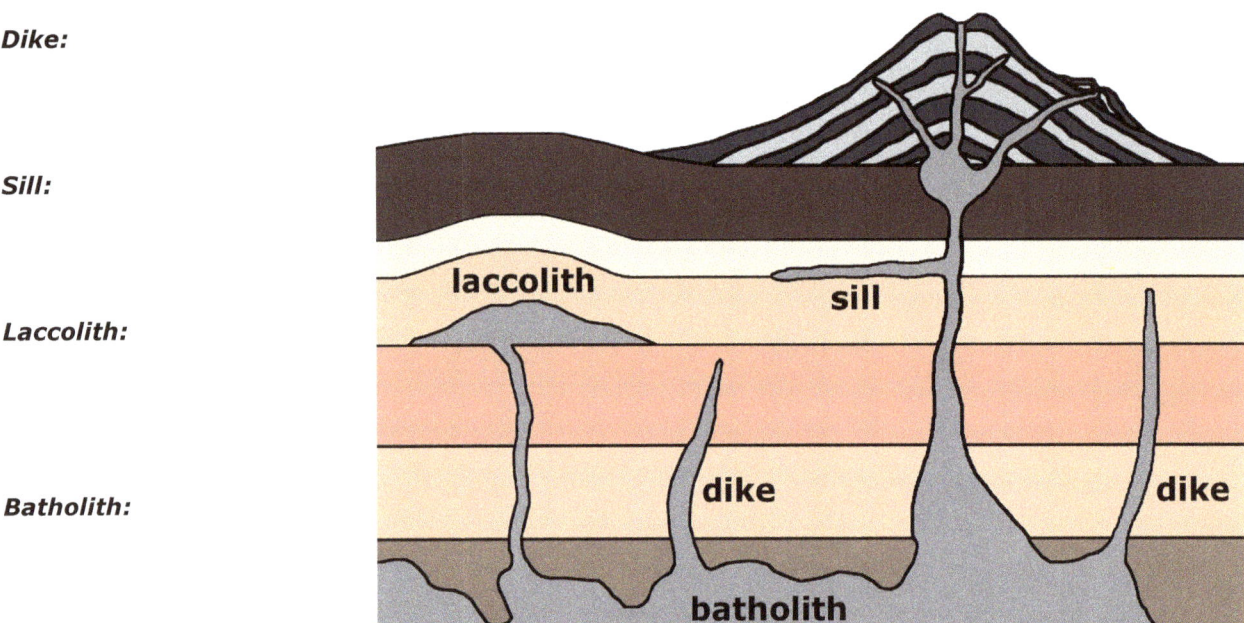

The different plutons, all of which form underground.

Also, note that intrusive igneous rock tends to be significantly more resistant to weathering than surrounding rocks. So, over time, they may become exposed as prominent features, as less-resistant rock is preferentially removed from around them. If this occurs and a batholith is exposed at the surface, it is called a **dome**.

Examples:

Stone Mountain in Georgia is an exposed batholith, and is thus a dome.

Half Dome in Yosemite National Park, California, is another exposed batholith. About half the batholith is missing, probably due to glacial activity in the past.

The final feature is a **volcanic neck**, which is the erosional remnant of a volcano. A volcano is not very resistant to weathering and erosion, with the exception of the intrusive igneous rock that formed from the last magma present in its pipe. So, in essentially the same way a dome is formed, the former pipe may become prominently exposed over time, and is then called a neck.

Examples:

Devil's Tower in Wyoming (L) and Ship Rock in New Mexico (R) are volcanic necks.

Section 3.4: Mountains

A **mountain** is a large mass of earth and/or rock, rising above the level of the adjacent land. They can be produced by several processes and differ significantly in their size, overall form, and internal structure, with some being no more than piles of lava and others being geologically complex features. Most are the result of some form of tectonic activity, directly or indirectly, which can cause the ductile and/or brittle deformation of crustal rocks, and also produce volcanoes.

Fold Mountains:

Fold (or folded) mountains are created by **ductile deformation**, which is the bending or folding of layers of rock. When subsurface layers of rock under significant pressure from the weight of overlying rock are subjected to compression by being squeezed laterally, they may behave like a plastic and actually bend rather than break. This typically produces two types of folds, which usually occur together. **Anticlines** are upward folds, while **synclines** are downward folds.

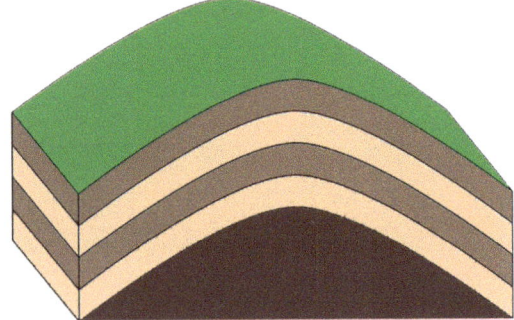

An anticline.

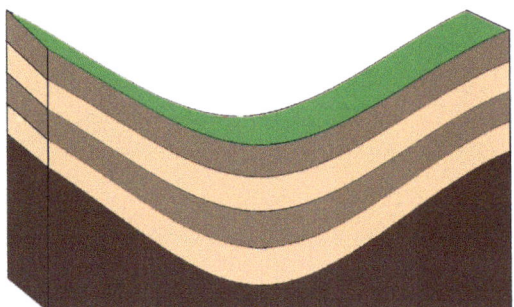

A syncline.

Fold mountains are typically formed at **continental-continental convergent plate boundaries**, where the collision of the two plates produces the lateral pressure required.

Still, because the subduction of oceanic crust and the formation of a volcanic island arc precedes a continental-continental collision, **fold mountains are the most geologically complex**.

Examples:

An anticline and syncline together.

Part of the folded Appalachians from space.

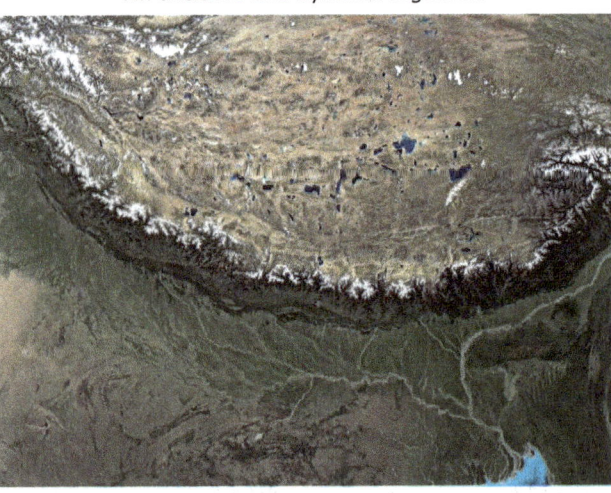

The snow-capped Himalayas from space.

Fault Block Mountains:

Fault block mountains are created by **brittle deformation**, which is the breaking and movement of layers of rock. When layers of rock are subjected to compression or tension, they may behave like a brittle solid and break rather than bend. This typically produces one of four types of **faults**, which are fractures in rock layers across which there has been significant movement.

When looking at these faults, their nature is determined by the relative movement that each side of the fault makes. The body of rock on either side of a fault is called either the hanging wall block or footwall block, and either can move up or down relative to the other. The **hanging wall block** is the one which leans over, while the **footwall block** is the one which slopes down.

Normal fault:

Reverse fault:

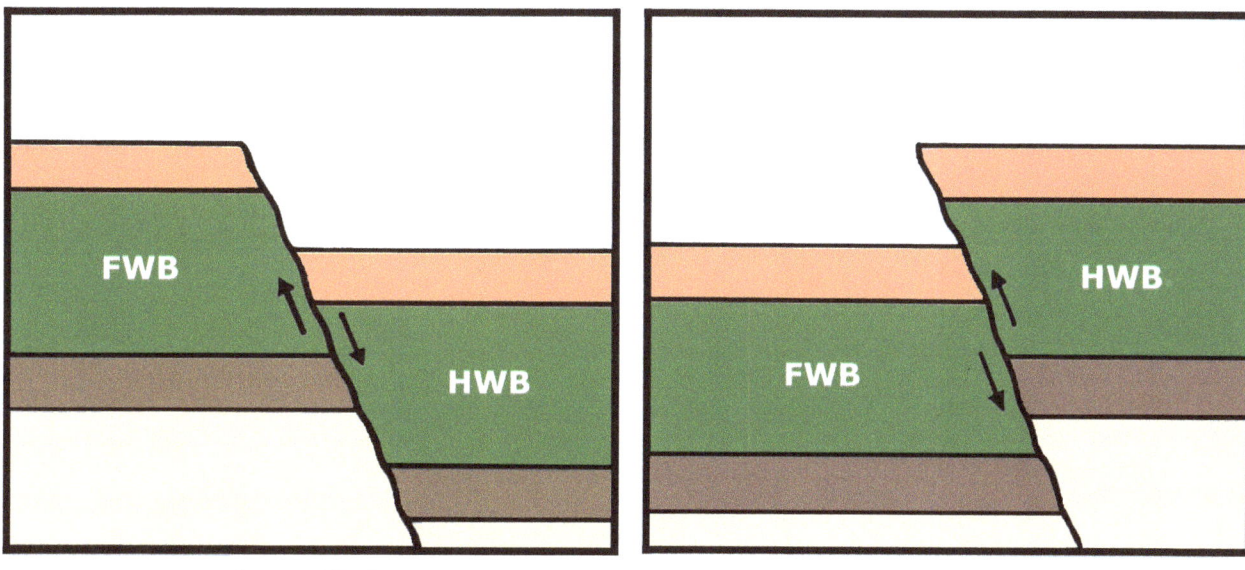

A normal fault.   A reverse fault.

Thrust fault:

Horst and Graben faulting:

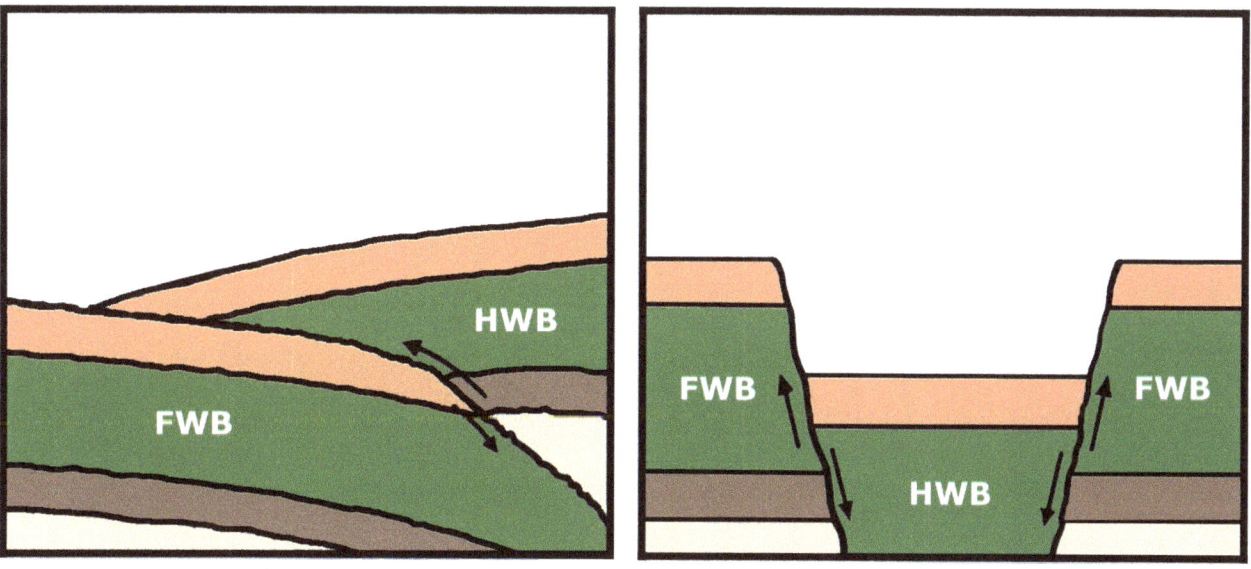

A thrust fault.   Horst and Graben faulting.

An exceptional example of an area dominated by fault block mountains is the **Basin and Range Province of the American Southwest**. Numerous near-parallel mountain ranges can be found there, with sediment-filled valleys in between. Alluvial fans are plentiful, as are playas.

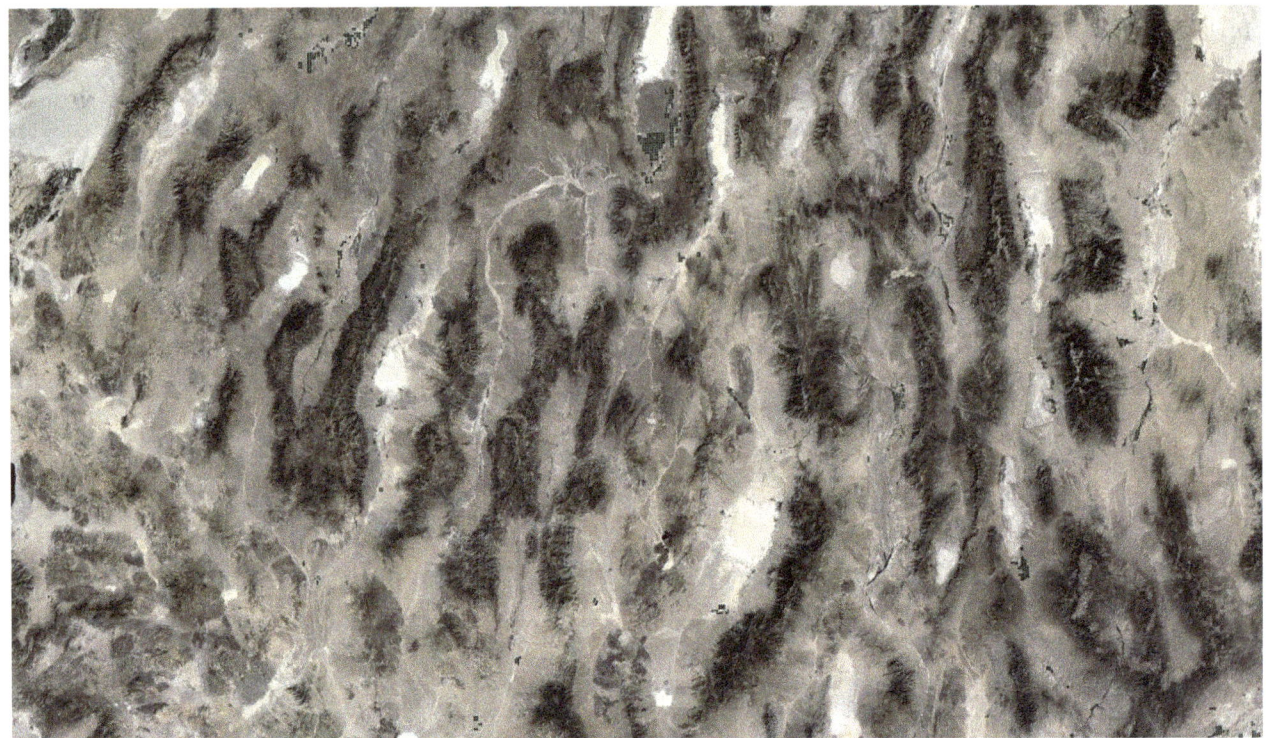

A portion of the Basin and Range Province, which covers most of Nevada and parts of surrounding states. Dark-colored mountain ranges can be seen, as well as valleys covered by alluvial fans and light-colored playas.

The province is dominated by mountains created by large-scale Horst and Graben and normal faulting.

Volcanic Mountains:

Obviously, volcanic activity can also create mountains, with the majority being found in **volcanic arcs** and **volcanic island arcs** created by the subduction of oceanic crust. Still, others may be created by magma rising from the mantle in areas not related to subduction, and sometimes far away from them.

Examples:

Every triangle is a volcano in the Aleutian Islands volcanic island arc.

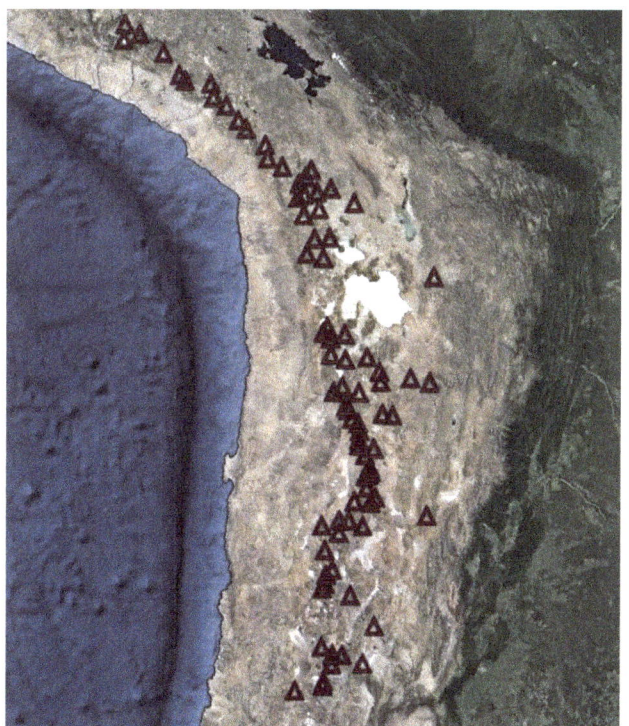
Numerous volcanic mountains in the Andes volcanic arc.

Volcanoes making up Hawaii Island, a volcanic hot spot.

Lastly, while technically not volcanic in origin, **domes** are mountains formed over time by the weathering and erosion of material from around a batholith, which is an igneous feature, nonetheless.

Examples:

On the right, a large un-named dome in the Libyan Desert that is approximately 12 miles across.

105

Unit 4

Section 4.1: Seawater, the Ocean Basins, and the Seafloor

Seawater covers most of the Earth's surface. In fact, seawater covers **approximately 71%** of it, which is why the Earth is often called the "Blue Planet". This water is salty because it contains significant quantities of dissolved solids with the predominant one being salt, of course, which is composed of sodium and chlorine. However, sulfur, magnesium, calcium, and potassium are also present in significant quantities, with many other elements also being present in very low concentrations. While it varies to some degree from place to place, all together these make up **about 3.5%** of a given volume of seawater.

These dissolved solids primarily come from two sources:

Almost all of the Earth's water is held in its five ocean basins, which are the Atlantic, Indian, Southern, Arctic, and Pacific Oceans. A significant amount is also found in its many seas, such as the Caribbean and Mediterranean, and in the Gulf of Mexico.

The depths of the oceans varies greatly, as they range from sea level to the bottoms of trenches created by the subduction of oceanic crust. Still, if averaged out, the global average depth of the oceans is about **12,000 feet**, or about **2.3 miles**.

When it comes to the deepest location on Earth, 2.3 miles is rather shallow, though. This location is called the **Challenger Deep**, and is in the **Mariana Trench** in the Western Pacific. It's a remarkable 35,814 feet to the bottom there, which is **almost 7 miles**. That's about as far down as large jetliners fly up, and only three people have ever been there. The water pressure is about 1,230 bars, whereas average atmospheric pressure is about 1 bar. So, it was only possible in two specially built submarines.

Within the ocean basins, the seafloor is divided into three major units, which are **continental margins**, the **ocean basin floors**, and **oceanic ridges**. And, when looking specifically at continental margins, it can be seen that there are two basic types.

#1 Active continental margins:

#2 Passive continental margins:

Passive margins can then be sub-divided into three basic parts, as well.

#1 Continental shelf:

#2 Continental slope:

#3 Continental rise:

The rise is comprised of thick layers of sediment from land, which form **abyssal fans**, and are something like underwater alluvial fans. These fans are created by the downward movements of sediments off the shelf and down the slope, called **turbidity currents**, which are like underwater avalanches. These are typically initiated by an over-accumulation of sediment at the edge of the shelf or by earthquakes.

Conversely, active margins **have a trench offshore, and tend to have a very narrow shelf**, if any. This is because most sediment from land moves into the trench and is subducted along with the oceanic plate, and thus does not accumulate along the edge of the continent.

The location of the Challenger Deep, part of the Mariana Trench, which is in the West Pacific.

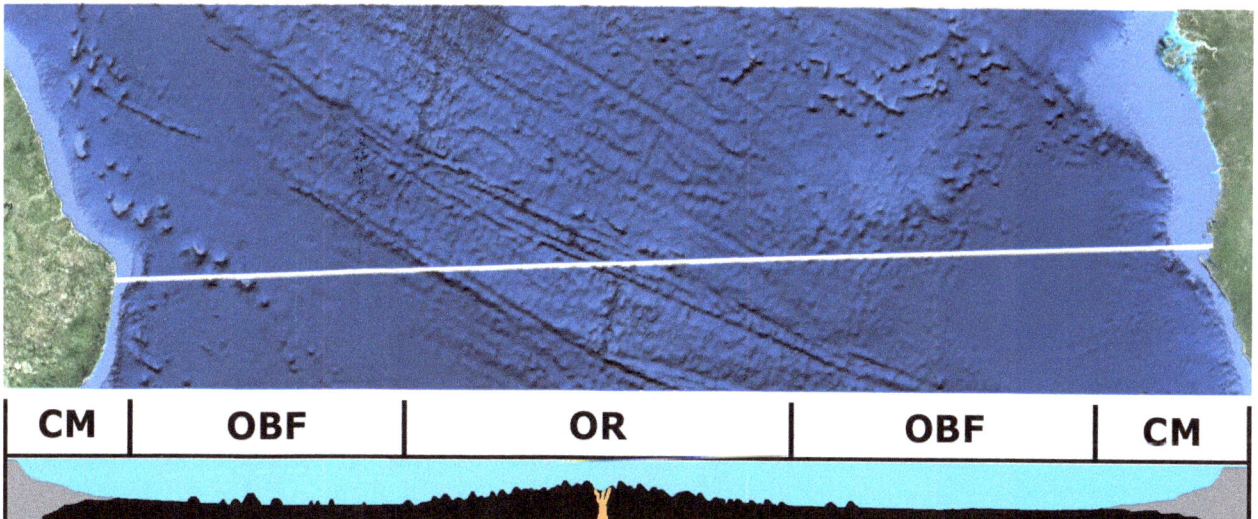

A cross-section across the South Atlantic, showing seaward sloping continental margins (CM), the relatively flat ocean basin floors (OBF) called abyssal plains, and the elevated oceanic ridge (OR).

South America is a good place to see distinctive active and passive continental margins. On the west coast is an active margin, with a relatively small shelf and offshore trench created by subduction. On the east coast is a passive margin, with a relatively large shelf and no trench.

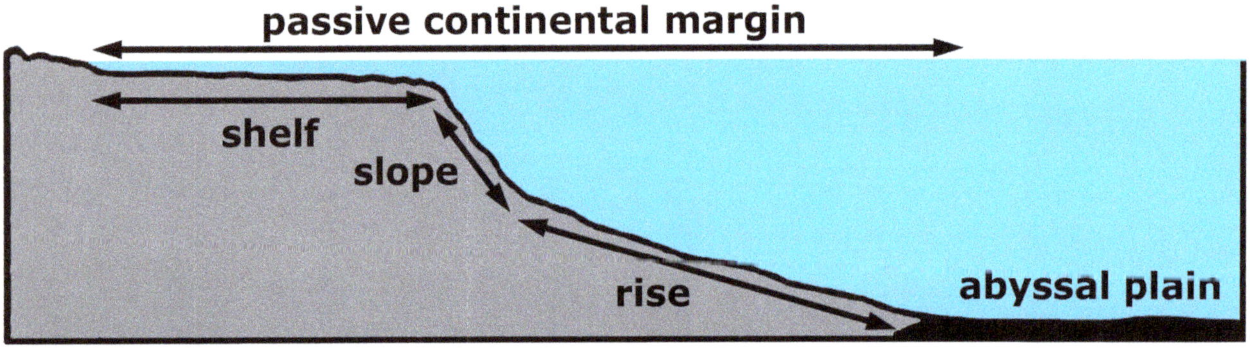

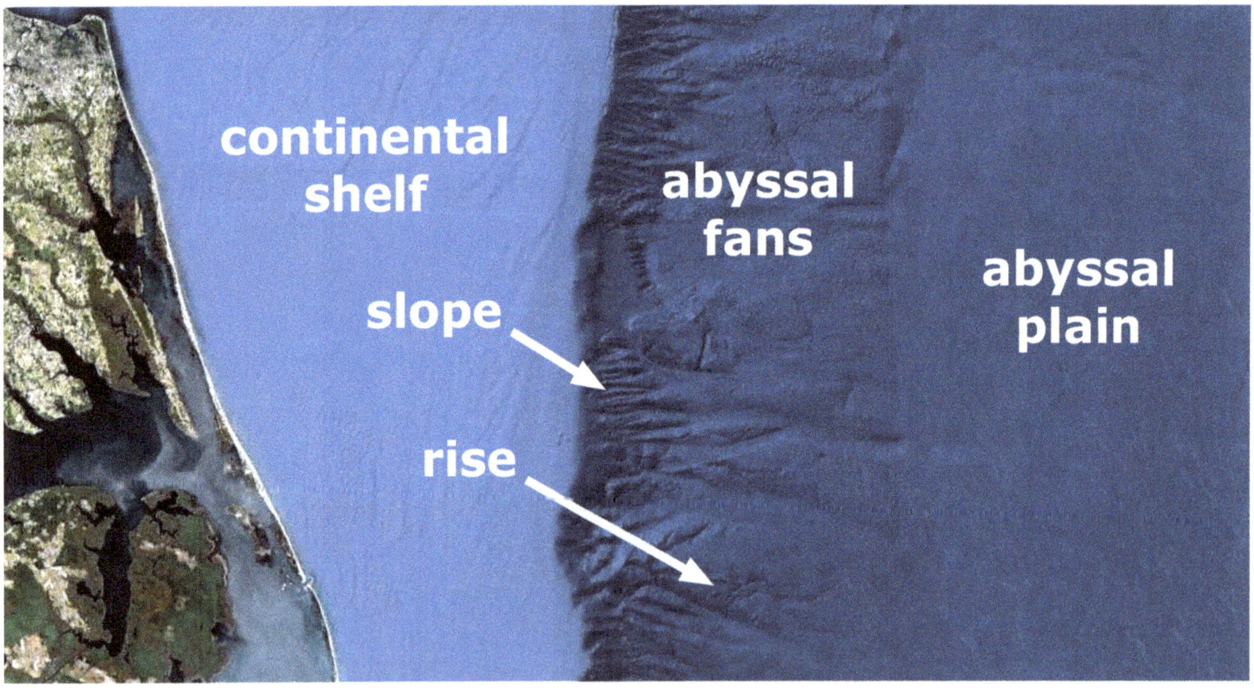

The three basic parts of a continental margin, which are the shelf, slope, and rise, and the deep abyssal plain.

Most of the ocean basin floors are called **abyssal plains**, which are the relatively flat and featureless parts of it. However, scattered about the ocean basin floor, there are also numerous volcanic **oceanic islands** and **seamounts** produced by magma plumes/hot spots and tectonic activity.

Oceanic island:

Seamount:

When an oceanic island or seamount is produced by a stationary body of magma, it eventually stops erupting as the body of magma becomes depleted, or the plate on which the volcano formed slowly carries it away from the magma over time. Growth of the volcano then stops, and oceanic islands are subsequently subjected to continuous weathering and erosion by waves.

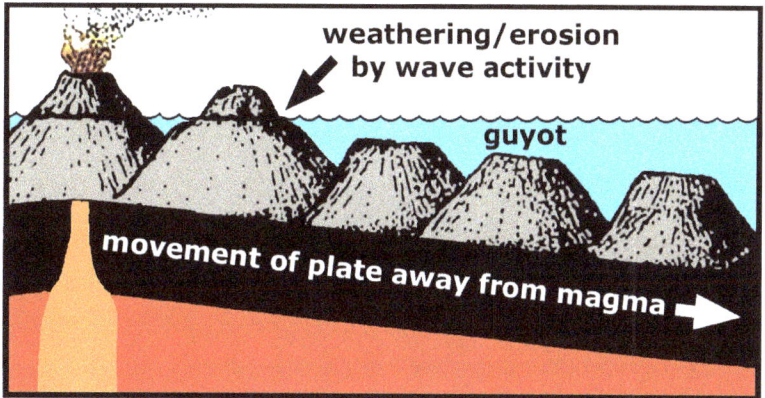

Because volcanoes poorly resist weathering, the top part of an oceanic island emerging from the water can be stripped away relatively quickly. Over time, the crust underneath an oceanic island or seamount also begins to cool, contract, and also increase in density, causing them to sink. So, an oceanic island can become a **guyot**, which is a flat-topped submerged volcano.

Also, in warm tropical waters, this process can lead to the formation of **atolls**, which are roughly circular coral reefs produced when the top of an oceanic island drops below sea level.

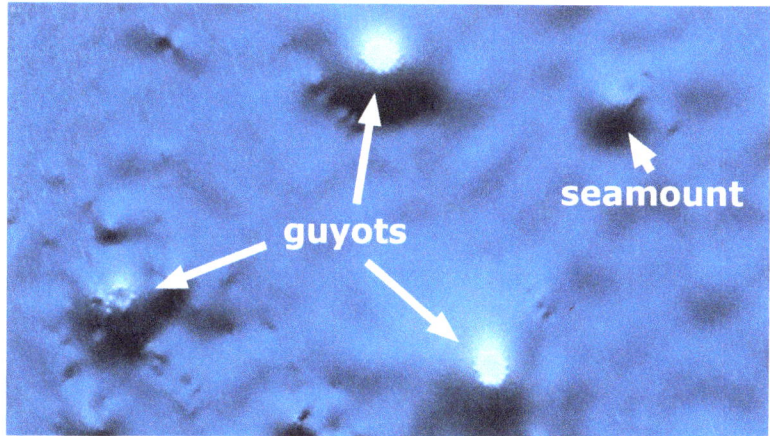

In this case, corals form a ring-shaped reef around an oceanic island, and as the island sinks over time, the reef grows upwards. Because the reef can grow up faster than the volcano sinks down, it can maintain its overall form after the island becomes submerged.

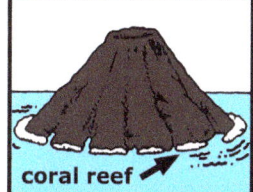  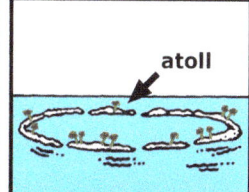

Waves, especially during storms, also break up the top part of reefs and can pile up broken pieces of coral, volcanic rocks, and sediments on top of the reefs. Thus, over time, part of a reef may become land. Vegetation can then take hold as floating seeds and those carried by seabirds accumulate on the newly-created land, and some large atolls are even inhabited by humans.

When corals spawn, their eggs can spread out for many days, and can produce new reefs in distant areas.

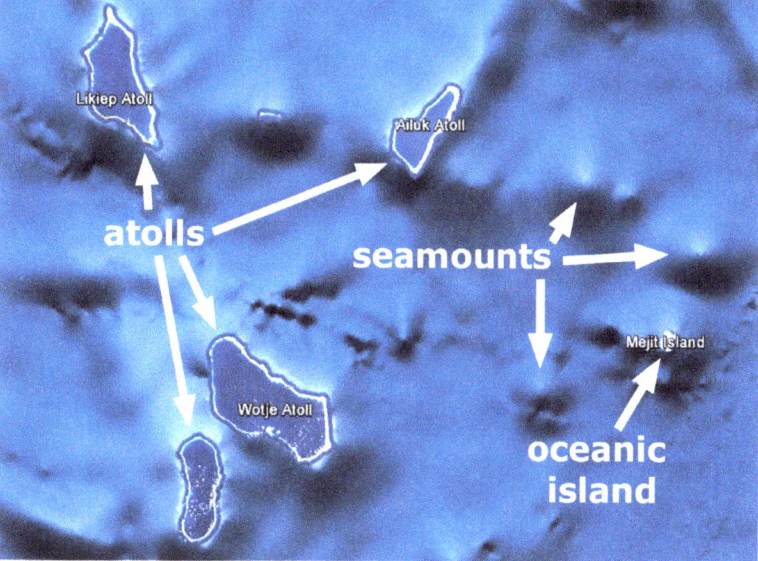

The formation of an atoll occurs when an oceanic island surrounded by a coral reef sinks below the surface.

Oceanic ridges are essentially underwater mountain ranges produced at divergent plate boundaries, which rise above the abyssal plains because they are comprised of relatively young and warm oceanic crust that is lower in density than the older, cooler crust farther from the boundary. **Rift valleys** are typically present at their center, and the ridge itself is comprised of **new oceanic crust** and **volcanoes**.

In the Atlantic, the oceanic ridge runs down the middle of the ocean basin, and is thus called a mid-ocean ridge, specifically the **Mid-Atlantic Ridge**. Those that do not actually run through the middle of a basin are still called mid-oceanic ridges, but some are named differently. For example, much of the Pacific Plate is created along the East Pacific Rise, while much of the Antarctic Plate is created along the Southwest Indian Ridge. Regardless, these ridges are many thousands of miles long, and wrap around the Earth much like the stitches on a baseball.

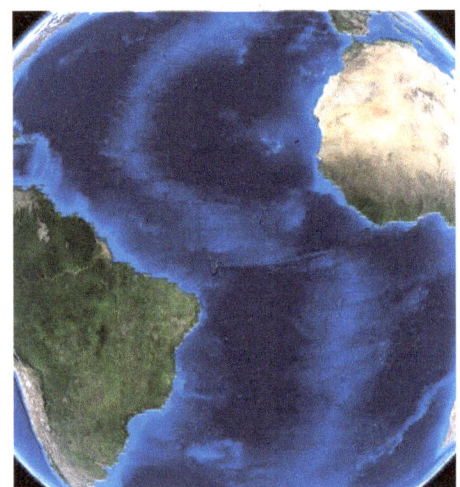

The Mid-Atlantic Ridge.

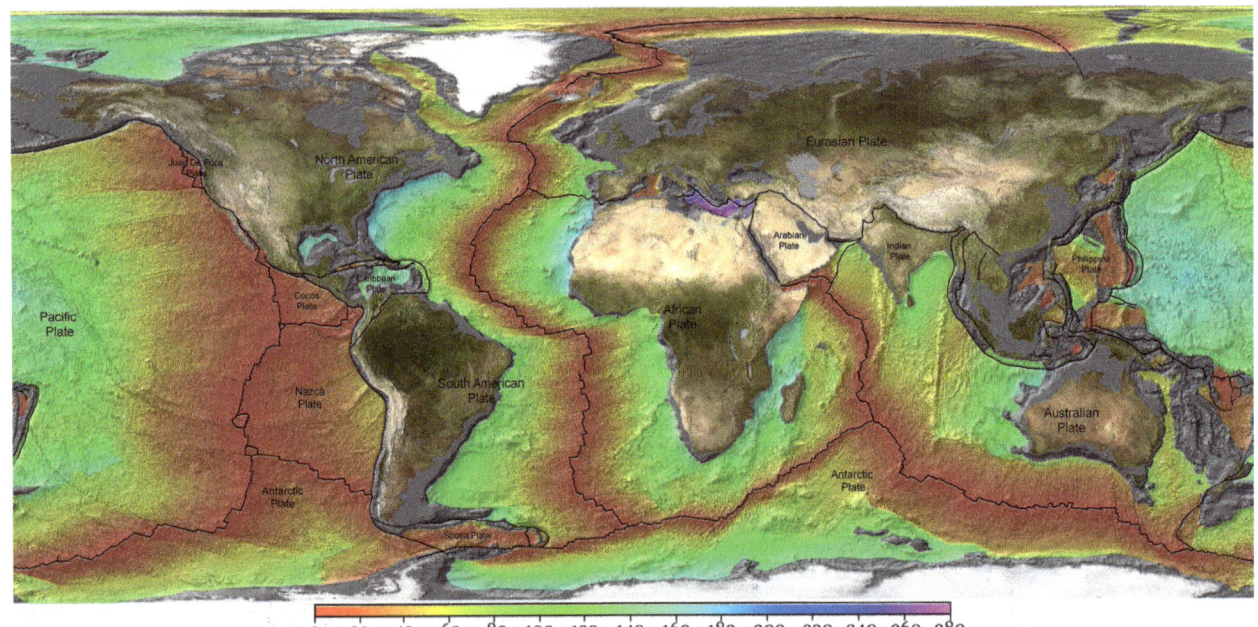

The age of oceanic crust in millions of years is shown, and the mid-oceanic ridges can be seen as black lines in the ocean basins, around which the crust is new.

Sediments of various sorts blanket the entirety of the ocean basins, with some coming from land and some coming from other sources. Those originating from land are called **terrigenous sediments**, and tremendous quantities are carried to the oceans by streams. Much of this sediment is relatively coarse-grained and heavy though, and thus settles out of the water relatively close to land on/near continental margins. However, fine-grained terrigenous sediments can be carried out farther from land, primarily by currents and winds, and slowly sink to the bottom.

Regardless, in deep waters far from land, the seafloor is covered by **pelagic sediments**, which are all fine grained. These consist **primarily of a mixture of fine-grained terrigenous sediments and the tiny shells of dead plankton**, which are called **biogenous sediments** since they are produced by living things. Pelagic sediments may contain **volcanic ash** and **meteoric dust** that enters the atmosphere from space, as well.

Lastly, some pelagic sediments are called **oozes** if they are comprised of **at least 30% biogenous sediment**, and these oozes cover about 2/3 of the seafloor. Pelagic sediments in the remaining areas contain **less than 30% biogenous sediment** and are called **pelagic clays**.

Streams transport terrigenous sediments to the oceans.

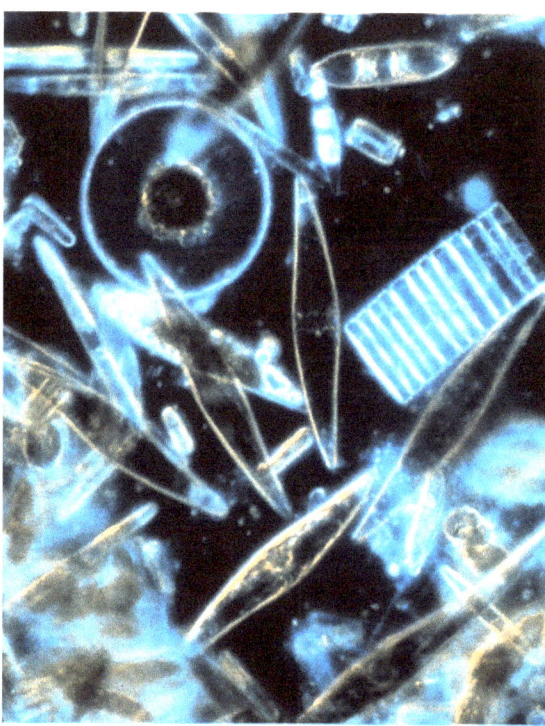
Biogenic sediments are primarily dead plankton.

Sediments entering the oceans may also come from space, volcanoes, and dust storms.
Above is a very large dust storm moving from West Africa out over the Atlantic Ocean.

Section 4.2: Ocean Circulation and the Tides

Ocean currents are masses of water that move from one place to another, and water perpetually flows from one place to another in all of the Earth's major bodies of water. Large currents act as conveyor belts that move water, both shallow and deep, and the entire global ocean circulates due to the fact that all of the oceans are connected to each other. **Tides**, on the other hand, are periodic changes in local sea level created by the gravitational pull of the Moon and the Sun.

Ocean Circulation:

There are surface currents, deep ocean currents, and vertical currents circulating water throughout the oceans. These currents are simple in some ways, yet quite complicated in others, as several different mechanisms create and affect them. The flow of water is interrupted to some degree by the continents, large islands, oceanic ridges, etc., as well.

Surface currents are created and driven by winds, due to the friction generated by the air moving over the water's surface. These move water laterally across the oceans' surfaces, down to a depth of about 1,300 feet.

Like airflow, the movement of water at the surface is also acted upon by the Coriolis Effect. So, as water is moved by wind, it is typically deflected to the right or left, depending on the hemisphere. This creates numerous loop-like surface currents, which are called **gyres**, and some of these are quite large in size.

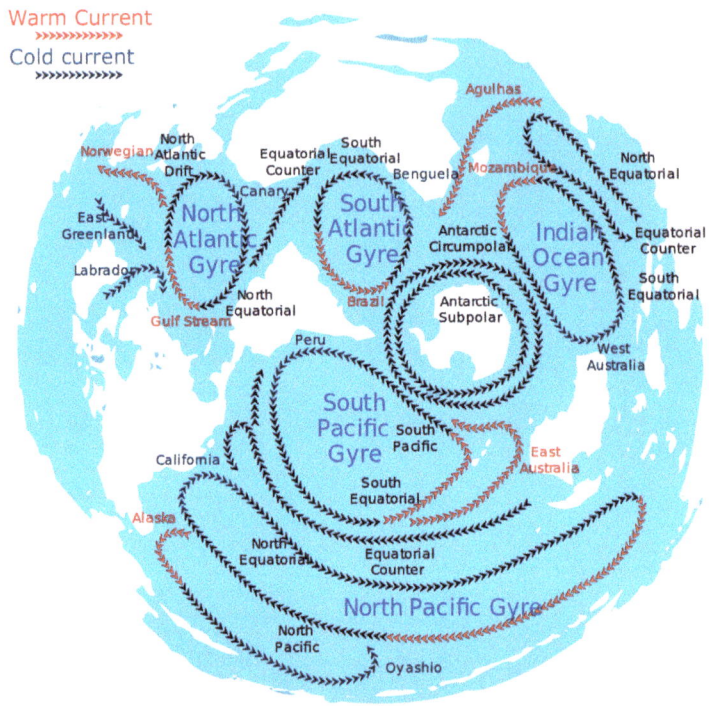

The major gyres, as seen from the south.

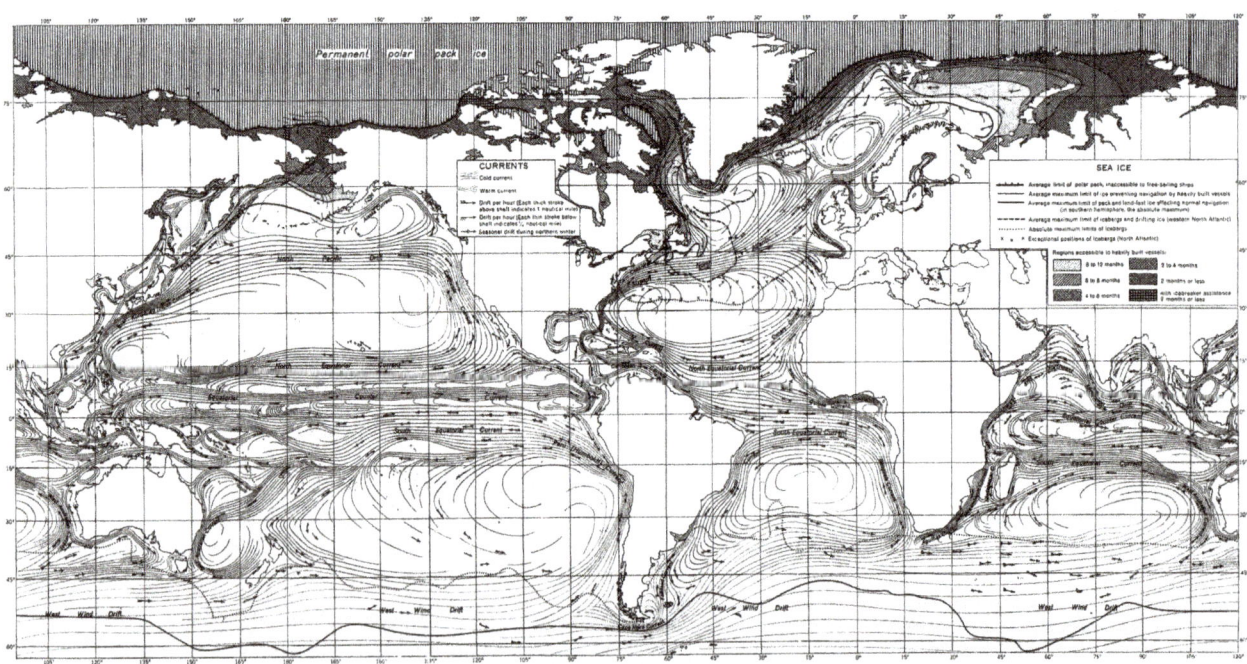

A military map from the 1940's showing the major gyres.

112

One of these, the North Atlantic Gyre, moves water around the North Atlantic in a clockwise fashion under the influence of the prevailing winds, a notable part of which is called the **Gulf Stream**. This warm water current starts in the Gulf of Mexico, curves past South Florida, and then moves up the U.S. East Coast. It also affects temperatures on nearby land throughout the year.

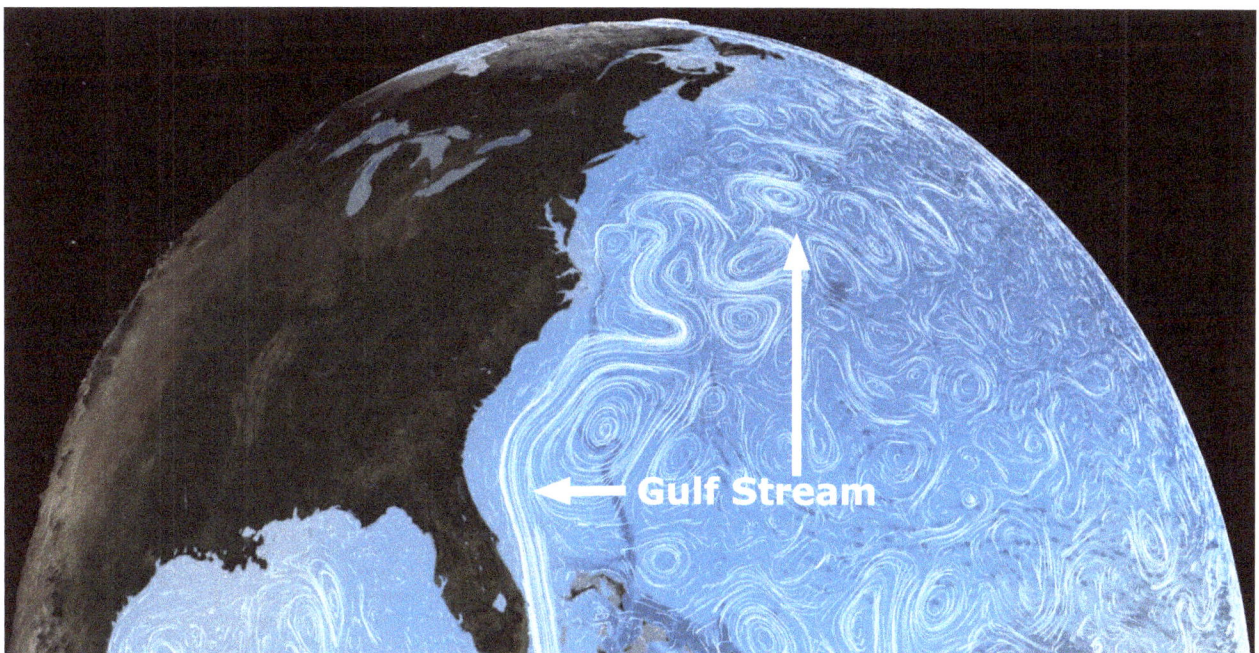

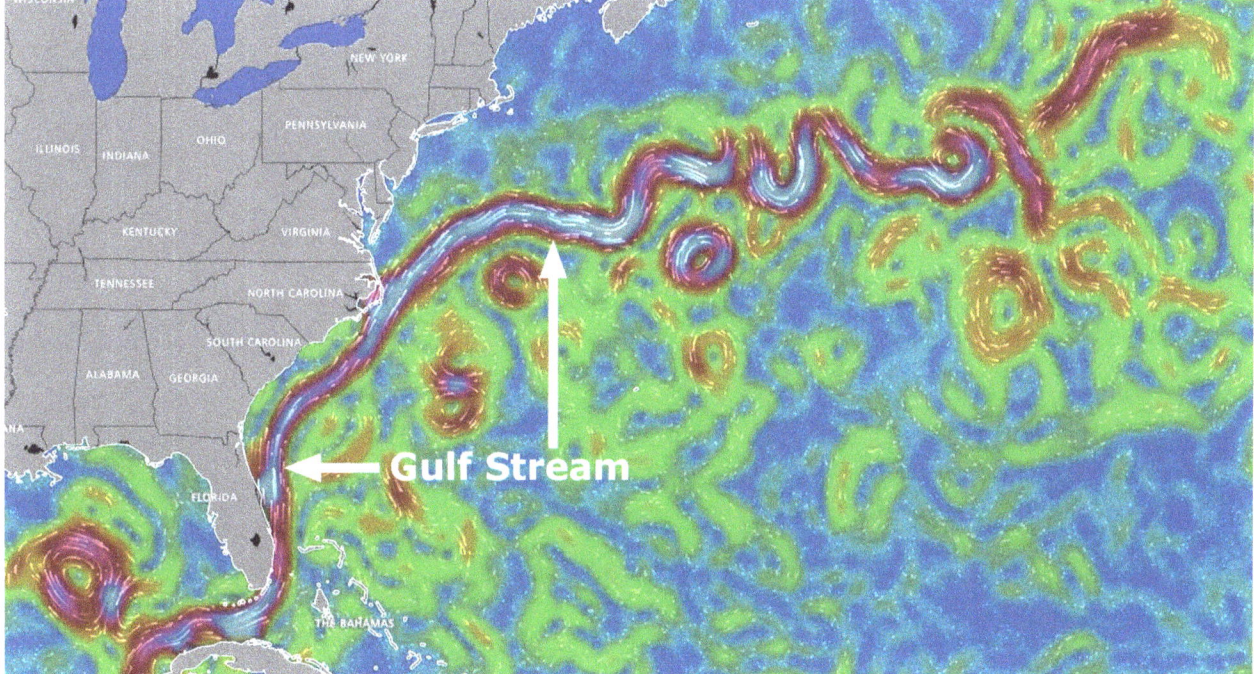

The Gulf Stream is clearly visible when looking at water motion (top), and especially when looking at water temperatures (bottom). Note how complex surface currents are, as well.

As water moves toward the poles, it becomes more dense and sinks to the seafloor. From there, it flows across the bottom of the oceans and creates **deep ocean currents**. The creation and flow of these currents is called **thermohaline circulation**, and is the result of **changes in water temperature and salinity, and thus density**.

Several things come into play with respect to the density of water. For example, **where precipitation is frequent, or where a large river discharges freshwater into an ocean**, the seawater becomes less salty, and thus less dense, while **high rates of evaporation** in some areas make seawater even saltier, and thus more dense. Likewise, if seawater is warmed it expands and becomes less dense, while it contracts and becomes more dense if it is cooled. And lastly, **when floating sea ice forms at the surface of the Arctic and Southern Oceans,** the dissolved salt in the freezing water is rejected as the ice forms. This leaves the seawater just underneath the ice even saltier and thus more dense, while seawater becomes less salty and less dense where sea ice melts. So, as surface currents carry seawater to the polar regions, it **cools** and its density increases, and it becomes even more dense if its **salinity increases** where sea ice is forming. This is why it sinks and creates deep ocean currents.

Salinity and density increase where sea ice forms.

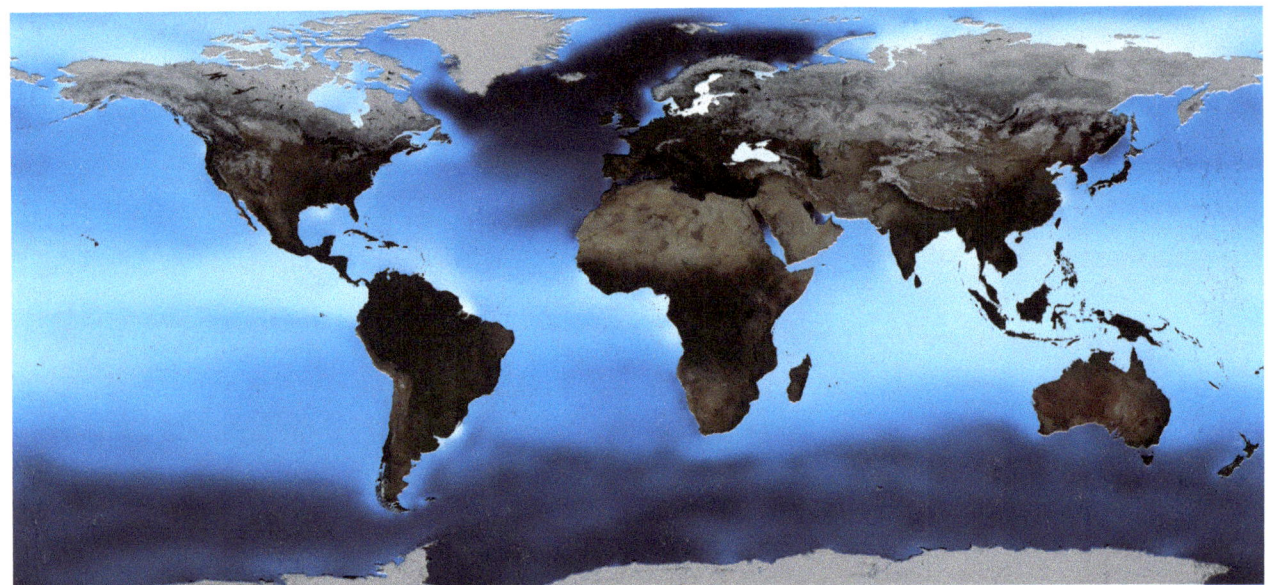
Average surface water density is shown, with light blue being less dense and dark blue being more dense. Note the elevated density of seawater near Greenland and the Arctic Ocean, as well as around Antarctica.

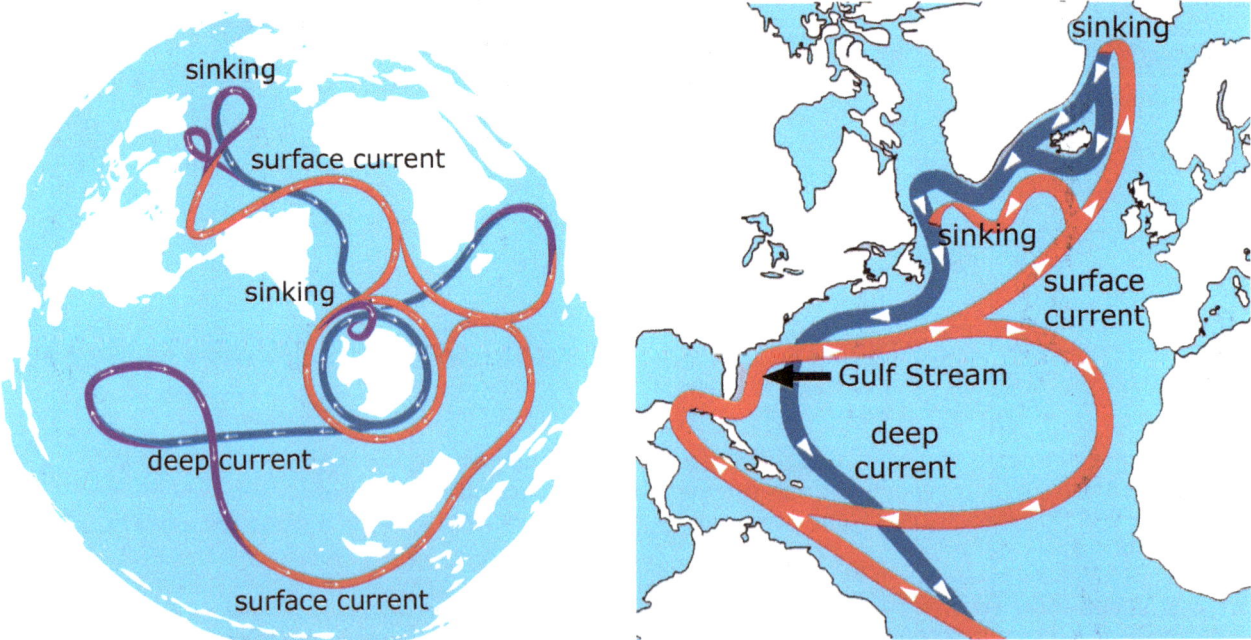
The major deep-sea currents are shown, which are created and driven by thermohaline circulation.

If seawater sinks in some areas and moves away via deep ocean currents, it must also rise back to the surface in other areas. When this occurs, it is called **upwelling**.

Seawater at depth moves up to the surface via upwelling **where winds, in combination with the Coriolis Effect, cause surface water to move away from continents**. Then, as the surface water moves away, it draws up and is replaced by water from underneath.

While the Coriolis Effect is weakest at the equator, and most surface water moves westward, winds can still cause some surface water to flow poleward. So, as it moves away from the equator, it is also replaced by seawater from underneath, which is called **equatorial upwelling**.

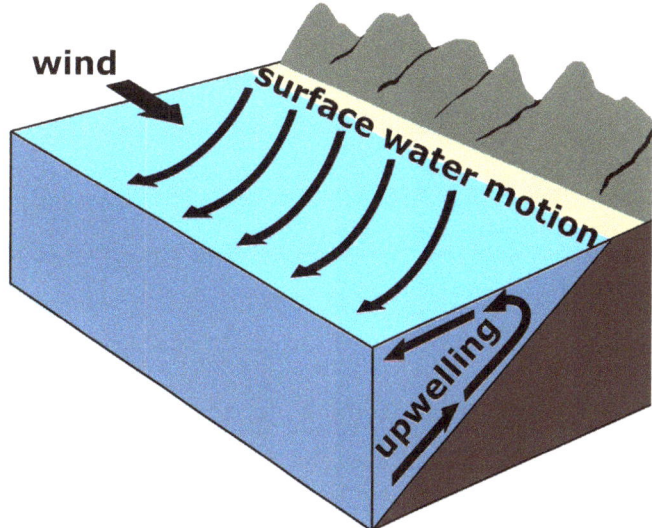

An example of an area where wind creates upwelling.

As a whole the combination of surface currents, deep ocean currents, and upwelling create continuous planet-wide water circulation, which is called the **Global Ocean Conveyor Belt**.

Lastly, the trade winds blow from east to west in the tropics in normal years, and move warm surface water away from South America and across the Pacific toward Indonesia and Australia. There, it literally piles up into an enormous mound of warm water that increases sea level by as much as two feet in the region. As this warm surface water moves away from South America, it also fosters the upwelling of cooler water from below.

This cool upwelled seawater cools the air off the west coast of South America, and thus reduces humidity, cloud formation, and precipitation, while the warm seawater moving westward warms the air in the West Pacific, and thus increases humidity, cloud formation, and precipitation there.

However, for unknown reasons, the trade winds weaken every 3 to 7 years, and this westward flow of water reverses. Then, the mound of warm water begins to flow back toward South America due to gravity. The change in flow reverses the normal weather patterns, causing the East Pacific to warm significantly and become wetter, while the West Pacific experiences drought. This is called an **El Niño**, and because conditions in the atmosphere are so strongly affected by the oceans, it can produce anomalous weather over much of the planet. Many areas are flooded while others dry out.

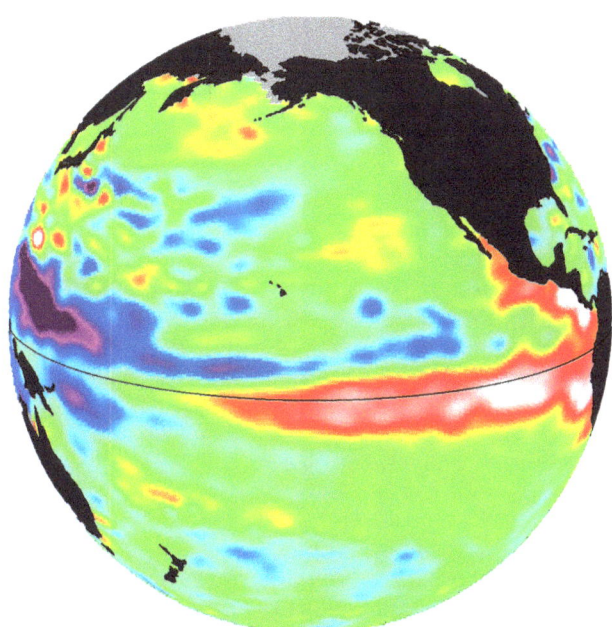

Surface water temperatures during an El Niño. Note the accumulation of cool water in the West Pacific, and warm water in the East Pacific and off the coast of South America. This is a reversal of the normal pattern, producing anomalous weather in many areas.

The Tides:

Again, the tides are periodic changes in local sea level created by the gravitational pull of the Moon and the Sun. They are caused by slight variations in these gravitational attractions, as the pull of the Moon and the Sun deform the Earth slightly, but can deform the fluid oceans to a much greater degree. As both pull on the oceans, tidal bulges are produced, which the Earth rotates under. Thus, when an area moves under a tidal bulge, local sea level rises, which is called a **flood tide**. And, when the area subsequently moves out from underneath the tidal bulge, sea level falls, which is called an **ebb tide**.

The Moon actually produces two tidal bulges on the Earth through the effects of gravitational attraction, though. At the location on the Earth closest to the Moon, water is pulled toward the Moon, while simultaneously on the opposite side of the Earth, another tidal bulge is produced directly away from the Moon. However, this second bulge forms because the pull of the Moon's gravity is at its weakest on the opposite side of the planet.

So, as the Earth rotates, coastal locations actually move underneath and out from underneath two bulges, causing most coasts to experience two high and two low tides daily, which is called a **semidiurnal tidal pattern**. In fact, if the Earth lacked landmasses, all locations would experience two equal high tides and two equal low tides daily.

However, because the movement of water is impeded by landmasses and the shapes of different basins, some locations will experience only one high tide and one low tide each day, which is called a **diurnal tidal pattern**. Likewise, some areas experience tides that are in between a diurnal pattern and a semidiurnal pattern, which are thus called **mixed semidiurnal** or **irregular tidal patterns**, with a high high, a low high, a high low, and a low low tide occurring each day.

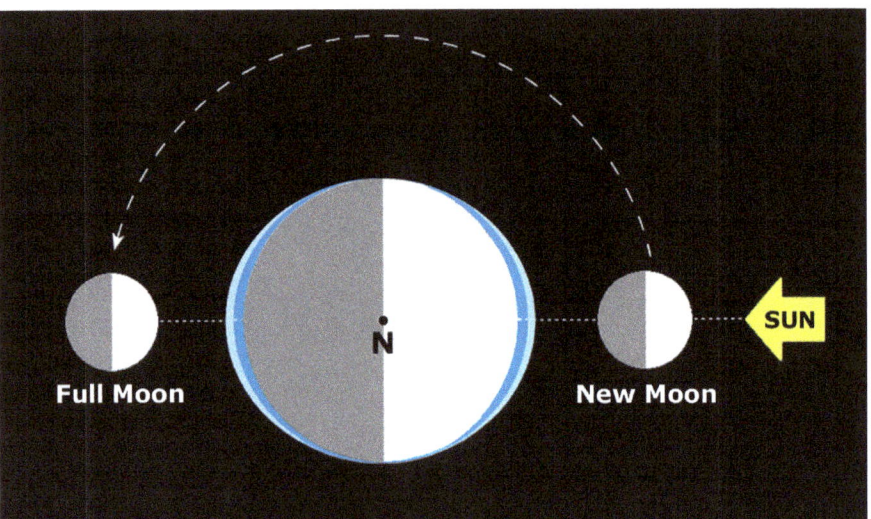

When the Earth, Moon, and Sun are aligned during the new and full phases of the lunar cycle, the lunar and solar bulges (light and dark blue) are compounded and create a spring tide. This is when the greatest tidal range is experienced.

The Moon revolves around the Earth too, making one orbit in about 27 days. So, because the Moon moves around the Earth, the tides do not occur at the same time every day for a given location. Instead, they occur about 50 minutes later each day. Thus, each full tidal cycle really takes about 24 hours and 50 minutes.

Because the Moon is much closer to the Earth than the Sun, it is the primary cause of the tides. But, the Sun's gravity still has a significant effect on the size of tidal bulges.

When the Sun is aligned with the Earth and the Moon, the combined effect of solar and lunar gravity enhances the tides. Thus, the highest high tide and the lowest low tide in a full tidal cycle are experienced on the same day, which is called a **spring tide**. This happens twice in a month's time, occurring during the **new and full phases** of the lunar cycle.

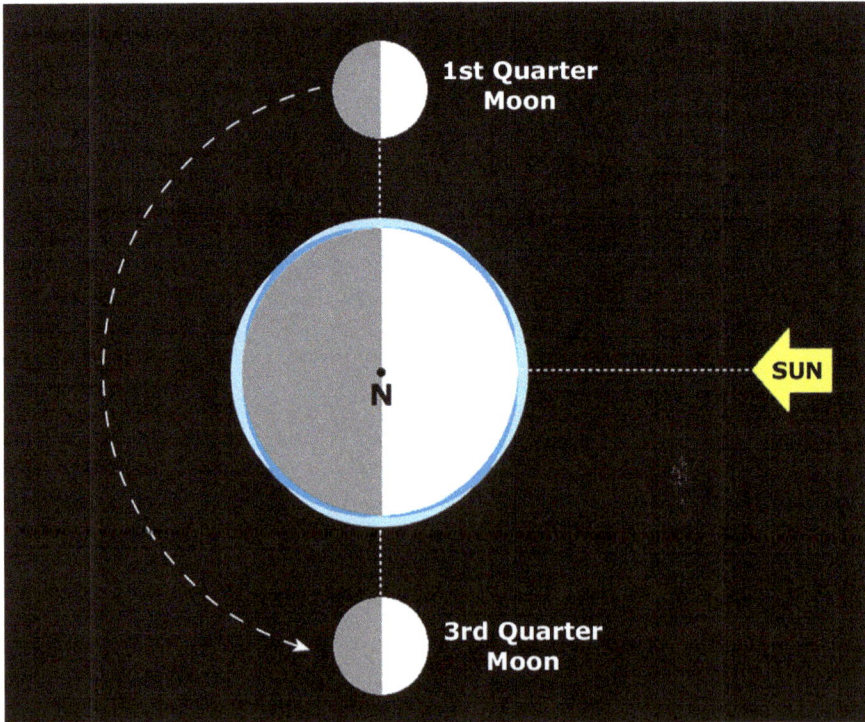

When the Moon and Sun are perpendicular to each other with respect to the Earth during the 1st and 3rd quarter phases of the lunar cycle, the lunar and solar bulges are opposed and create a neap tide. This is when the smallest tidal range is experienced.

On the other hand, when the Sun and the Moon are perpendicular to each other with respect to the Earth, the Sun's gravity counteracts some of the Moon's influence, which results in the smallest tidal range. Thus, the lowest high tide and the highest low tide are also experienced on the same day, which is called a **neap tide**.

Neap tides also occur twice in a month's time. However, they are experienced during the **1st and 3rd quarter phases** of the lunar cycle. Therefore, **spring and neap tides occur about a week apart.**

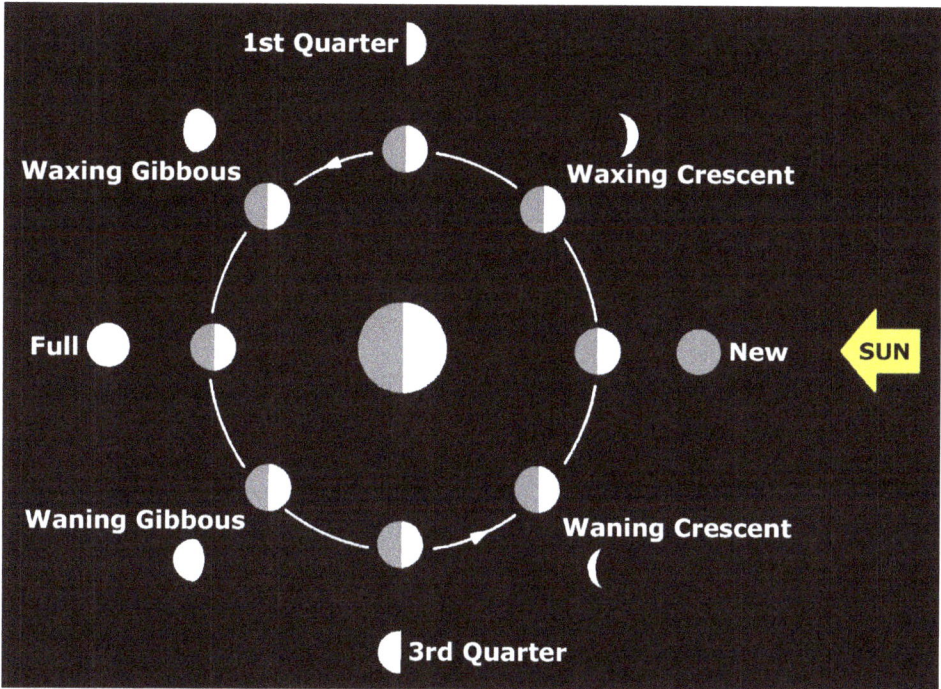

The phases of the Moon are shown. Note that the Earth, the Moon, and the Sun are aligned during the new and full phases, which creates spring tides, and that the Moon and the Sun are perpendicular to each other with respect to the Earth during the 1st and 3rd quarter phases, which creates neap tides.

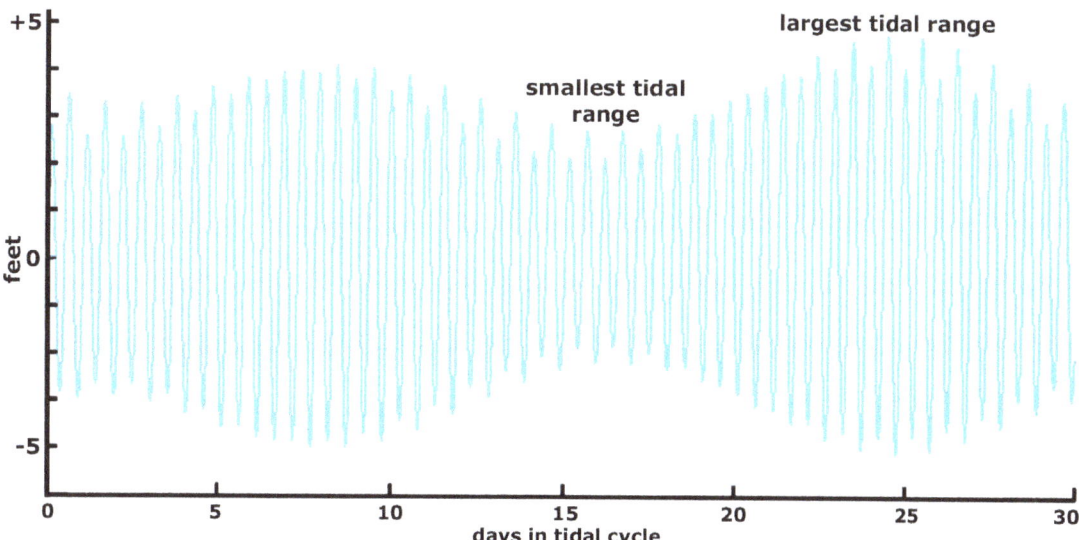

A typical tidal cycle over a period of about a month. Spring tides occur twice during the month and have the largest tidal range, and neap tides occur twice during the month and have the smallest tidal range.

Section 4.3: Coasts and Shoreline Features

Coastal areas are subjected to the elements, and wave activity in particular, which can modify them through weathering, erosion, and deposition. Waves can pound shorelines and weather and erode materials, and can also transport sediment supplied by streams that reach the ocean. This moving sediment can then be deposited where water movement decreases, and can produce a number of natural shoreline features. Human activities can also modify shorelines, as people build a variety of structures to control the movement of sediment and to protect shorelines from wave activity.

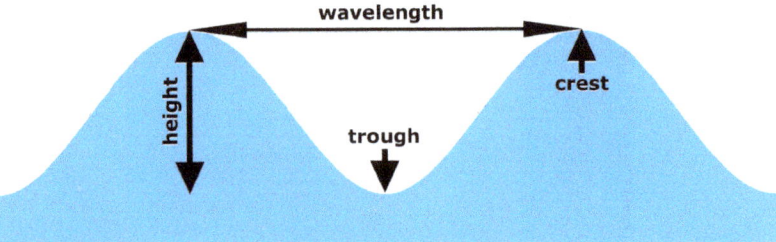

Looking at how waves work is the starting point for understanding coastal areas and shoreline features. So, the parts of a wave are shown to the right. The most important part is the **wavelength**, which is the horizontal distance from a wave crest to a crest or a wave trough to a trough.

In relatively deep water, waves do not interact with the seafloor, and the water in and underneath a wave is moved in a circular manner as a wave passes a given location. However, at a water depth equal to about **one half the wavelength**, which is called **wave base**, the wave will begin to interact with the seafloor. As the bottom of the wave contacts the seafloor, this creates drag, which slows the movement of the wave and leads to an increase in wave height with a corresponding decrease in wavelength. Then, the wave will break on the shore.

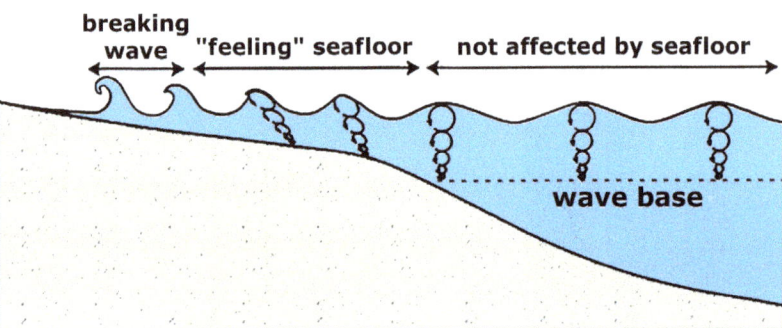

If a wave approaches the shoreline at an angle, it will begin to feel the seafloor as the first part of it reaches a depth equal to wave base. When this occurs, that part of the wave will slow down and the next part will catch up to it before slowing down, and so on. This has the effect of making the wave change shape and bend to the shoreline, which is called **wave refraction**.

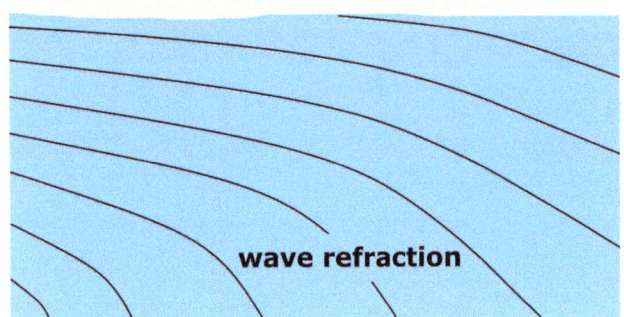

As refracting waves break onto the shore, sand is moved in a zig-zag path down the beach, which is called **beach drift**. Just offshore, wave refraction also leads to the movement of water parallel to the shore, called the **longshore current**, which also moves sediment. It is this movement of water, and of sediment by waves, that can lead to the formation of many shoreline features.

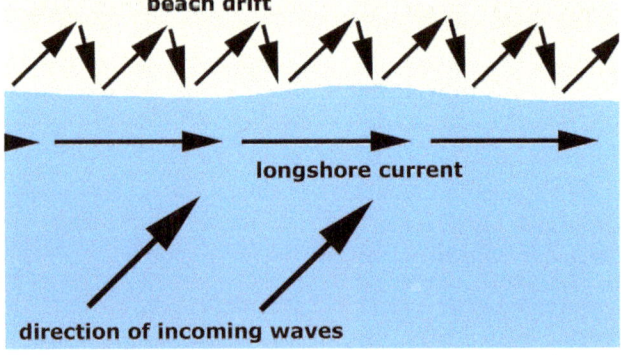

Common shoreline features that are produced in various coastal settings include:

Spits:

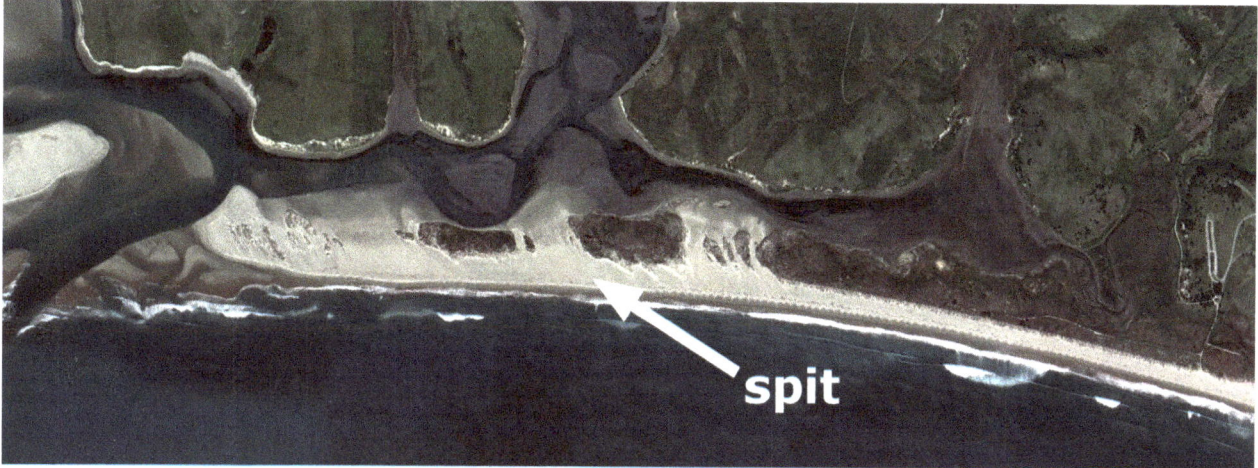

Baymouth bars:

Lagoons:

Tombolos:

Wave cuts:

Wave-cut cliffs:

Sea arches:

Sea stacks:

Beach dunes:

Barrier islands:

Common shoreline features that are produced by humans in various coastal settings include:

Breakwaters:

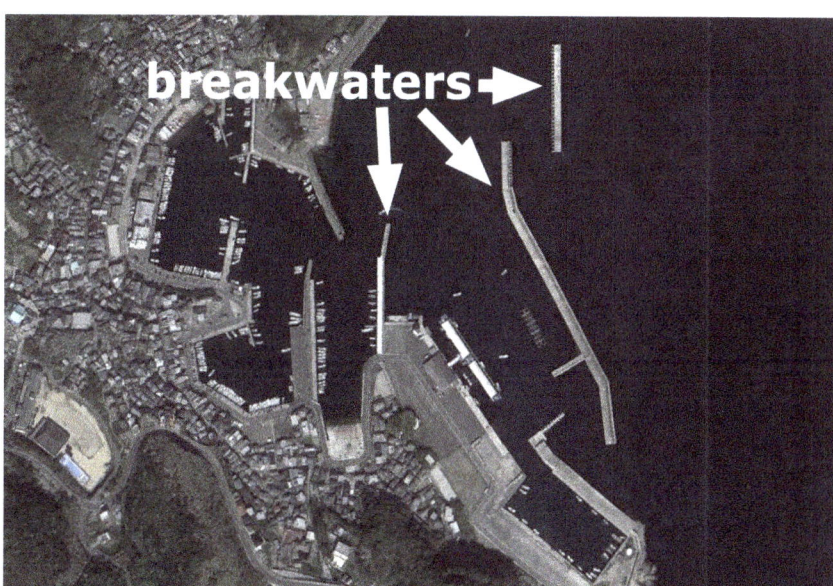

Seawalls:

Groins:

Jetties:

Lastly, humans can also simply build or build up a beach by collecting sand in a location where it is plentiful and transporting it elsewhere, typically using boats called dredge barges.

Oak Street Beach in Chicago, seen to the right, is a good example of such a beach, as it is man-made.

Many natural beaches in Florida are occasionally built up, or replenished, with transported sand to help battle coastal erosion, as well. Doing so is called **beach nourishment** or **beach replenishment**.

Section 4.4: Geologic Time

People long wondered how old the Earth was, and really didn't have any way of knowing until a new field of science, geology, was born in the late 1700's. As geologists and other scientists began to study the Earth, they found more and more evidence that it was far older than previously believed. It is hard for a person to comprehend how long even a million years is, and even harder to truly grasp how long 4.5 billions years is, or that the Earth is that old. Yet, centuries of study by many thousands of scientists have provided more than enough evidence to effectively prove that it is.

In addition to figuring out the age of the Earth, geologists and other scientists have built numerous histories of specific areas, and determined the timing of countless events that have occurred in the past. Two types of dating are used to do so, which are relative dating and absolute dating. **Relative dating** is the process of placing features and events in the correct sequence in which they formed or occurred. And, **absolute dating** is the process of determining the actual age of rocks and other materials.

Relative Dating:

Stratigraphy is the branch of geology that focuses on the study of **strata**, which are layers of rock, and the sequence in which they were formed. So, scientists in this field are the ones that develop the geologic histories of different areas, and collectively of the whole planet. To do so, they utilize several **Principles of Stratigraphy**, which are:

The Principle of Original Horizontality:

The Law of Superposition:

Strata in Grand Canyon are undeformed and horizontal. The oldest rocks are at the bottom of the canyon and decrease in age upwards.

The Principle of Cross-cutting Relationships:

The Principle of Inclusions:

Dike A cross-cuts sedimentary rocks, and thus must be younger than they are. Dike B cross-cuts dike A, and is thus even younger.

123

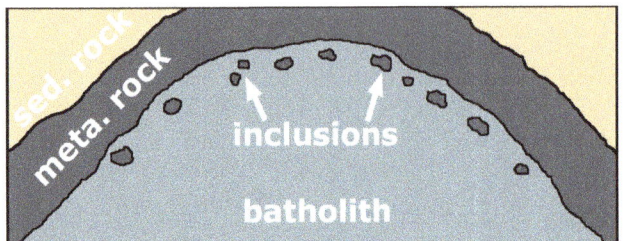

As a body of magma moves upward through other rocks it can heat and metamorphose them. Pieces of the surrounding rock can also become detached and get incorporated in the magma. Thus, it is common to find pieces of such rock trapped within the igneous rock that makes up a batholith. On the right is a piece of gneiss, a metamorphic rock, trapped within a granite boulder that was once part of a batholith. The gneiss is thus an inclusion, and is older than the granite.

The Law of Fossil Succession:

There were over 20,000 species of trilobite, each of which lived at a specific time. Thus, finding fossil trilobites of the same species in different rocks means the rocks are the same age.

In addition to the Principles of Stratigraphy, another important concept in stratigraphy is conformity. When sedimentary layers are deposited in a continuous manner, providing an uninterrupted record of geologic history, the layers are said to be conformable. However, changes in the geologic setting over time, particularly changes of sea level, can lead to the formation of **unconformities**. These are breaks in deposition, which typically involve the weathering and erosion of previously deposited material, as well. There are three types of unconformities, which are:

#1 Disconformities:

#2 Angular unconformities:

#3 Nonconformities:

A disconformity (A), angular unconformity (B), and nonconformity (C).

Strata in Grand Canyon are not conformable in many places, as millions of years sometimes separate a layer of rock from the layer over it. A disconformity is where these layers are parallel (left), and an angular unconformity is where they meet at an angle (at arrow in middle image). Likewise, in Death Valley nonconformities can be found where relatively young breccia, a sedimentary rock, is in direct contact with very old marble, a metamorphic rock (right).

Lastly, stratigraphers also use a technique called **correlation**, which is the matching of strata in different locations to determine the geologic history of an area. By looking at strata in the form of samples from boreholes (holes drilled into the ground for water, oil, etc.), and along roadsides, in canyons, etc., strata in one location can often be matched up with strata from another. This allows stratigraphers to piece together the unseen stratigraphy below the surface, and its extent, to form a more complete picture of the geologic history of the area.

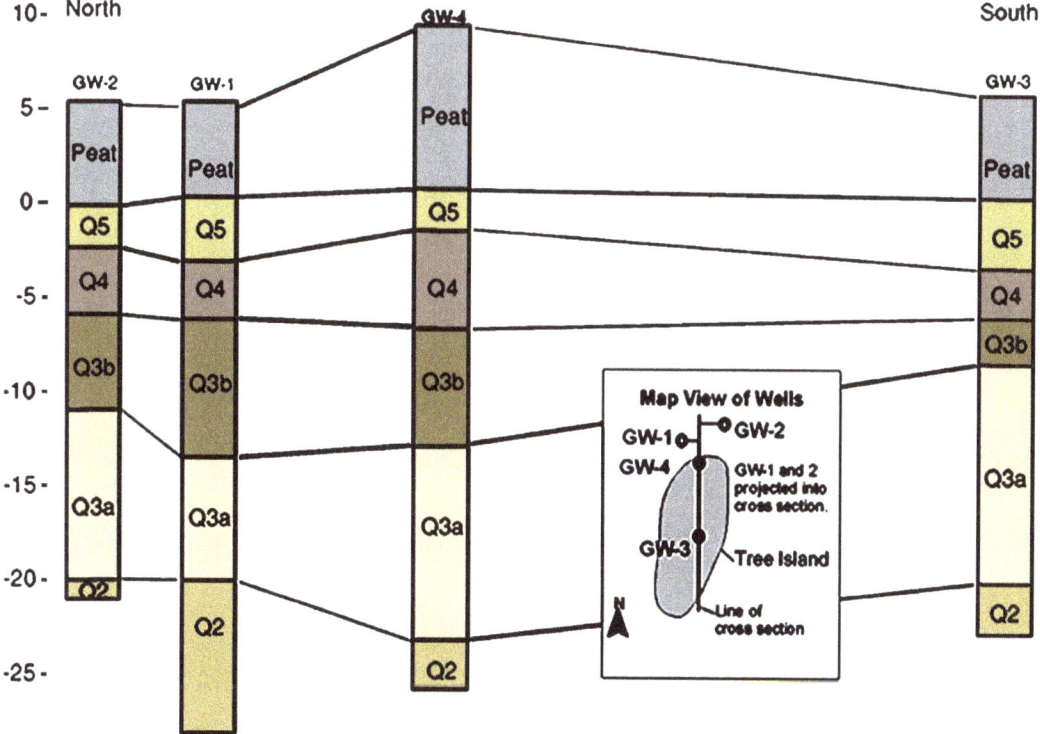

An example of correlation, using information from four well boreholes. The layers of sediment drilled through in each borehole can be matched up with those in the others to see the thickness and extent of each layer.

125

Absolute Dating:

Absolute dating, or radiometric dating, is accomplished by using the radioactive material found in some rocks to determine their actual age in years. Natural radioactivity involves the radioactive decay (breakdown) of the nucleus of an atom, and when this happens an original radioactive atom is transformed into a different form of the same element, or to a different element, altogether.

There are several such elements that can be used to radiometrically date rocks, and in each case the original radioactive element is called the **parent material**, which decays to a **daughter product** at a known rate. Since the rate of decay is known for each type of radioactive material, an absolute age can be determined by carefully measuring the amount of parent material and daughter product in a sample.

This is made possible because some mineral crystals will accept some parent materials into their structure as they form, but not the daughter product they decay to. So, a new crystal may have a significant amount of parent material incorporated into it, but no daughter product. Over time, the parent material slowly changes to its daughter product inside the crystal, though. Therefore, as the crystal ages, it will contain progressively less parent material and progressively more daughter product. So, the ratio of parent to daughter can be measured, and the age of the crystal can be calculated.

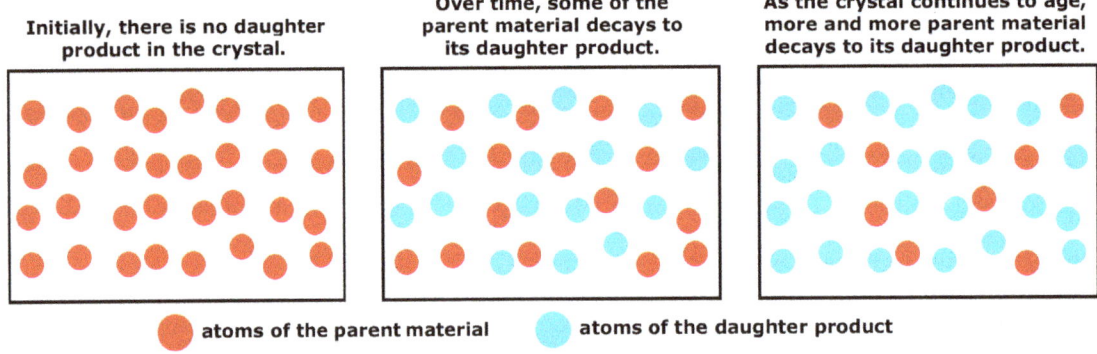

The Geologic Timescale:

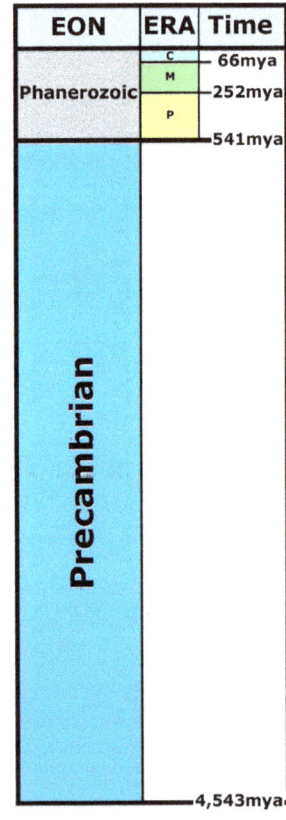

The Earth is approximately **4.5 billion years old**, and its history has been divided up into a number of chronological units, with the largest units being **eons**. The next smaller unit is an **era**, and eras are then subdivided into periods, epochs, and ages. Each division of Earth's history, regardless of length, has been made for a specific reason, with some notable occurrence or change taking place at that time. For example, the end-time of one era was chosen because it was the time at which the dinosaurs were wiped out. So, the divisions of the geologic timescale are not equal in length.

With that said, the Earth's history is split into two eons, which are the **Precambrian Eon** and the **Phanerozoic Eon**. However, **the Precambrian Eon spans almost 90% of the Earth's history**, with the rest being the Phanerozoic Eon.

The Precambrian Eon started when the Earth formed, about 4.5bya, and it ended about **540mya**. So, the Phanerozoic Eon spans the time from about 540mya until today. This time was chosen because many organisms began making hard parts, such as shells and spines, while before this time all organisms had soft bodies and no hard parts. This is significant because it is difficult for soft-bodied organisms to be preserved as fossils, and fossils more than 540 million years old are rare. On the other hand, the fossilization of various organisms' hard parts is far easier, and **fossils start to become quite common in rocks less than about 540 million years old**.

The Phanerozoic Eon is then subdivided into three eras, which are the Paleozoic Era, the Mesozoic Era, and the Cenozoic Era.

The **Paleozoic Era begins** when the Phanerozoic Eon begins, **540mya**, and **ends about 250mya**. At that time, the worst known mass-extinction event in the Earth's history occurred, which was actually far worse than the mass-extinction that killed off the dinosaurs.

The Paleozoic Era is also commonly referred to as **"The Age of the Invertebrates"**, which are animals that lack a backbone. However, some vertebrates, being the first fishes, amphibians, and reptiles did evolve during this era.

The **Mesozoic Era begins** when the Paleozoic Era ends, **about 250mya**, and **ends 66mya** with the extinction of the dinosaurs along with many other types of animals and plants. It is also commonly referred to as **"The Age of the Reptiles"** or **"The Age of the Dinosaurs"**.

Lastly, the **Cenozoic Era begins** when the Mesozoic Era ends, **about 65mya**, and **continues to today**. The first mammals actually evolved during the Mesozoic, but only became common after the extinction of the dinosaurs. So, the Cenozoic Era is commonly referred to as **"The Age of the Mammals"**.

Keeping all of this in mind, these names should make some sense. "Phaneros" means visible and "zoo" means animal. So, the Phanerozoic Eon is the time of visible animals (fossils). Likewise, "paleo" means ancient. So, the Paleozoic Era is the time of ancient animals. Likewise, "meso" means middle. So, the Mesozoic Era is the time of middle animals. And lastly, "ceno" means common. So, the Cenozoic Era is the time of common animals.

EON	ERA	PERIOD	Time
Phanerozoic	Cenozoic	Quaternary	2.6mya
		Tertiary	
			66mya
	Mesozoic	Cretaceous	
			145mya
		Jurassic	
			201mya
		Triassic	
			252mya
	Paleozoic	Permian	
			299mya
		Carboniferous	
			359mya
		Devonian	
			419mya
		Silurian	
			444mya
		Ordovician	
			485mya
		Cambrian	
			541mya

Section 4.5: A Brief History of Life

The Earth formed about 4.5 billion years ago, and the oceans were present by about 3.8bya. During this time the planet was bombarded by debris from space, which was still drifting about after the Earth's formation. However, the first signs of life appeared relatively quickly as conditions improved. In fact, **there is solid evidence that life arose in the oceans about 3.7 billion years ago**. At that time all living things would have been exceptionally simple organisms, similar to bacteria, though.

As time progressed, life evolved and diversified, yet for nearly 3 billion years, everything was still a single-celled organism. Then, about 900mya, the first multi-cellular life arose, with the first tiny animals appearing about 800mya. There was no significant life on land for another 300 million years, though.

In the Precambrian oceans, plants became commonplace and numerous types of invertebrate animals appeared, many of which were the ancestors of the organisms still present today. This was the Precambrian though, so everything still had a soft body,

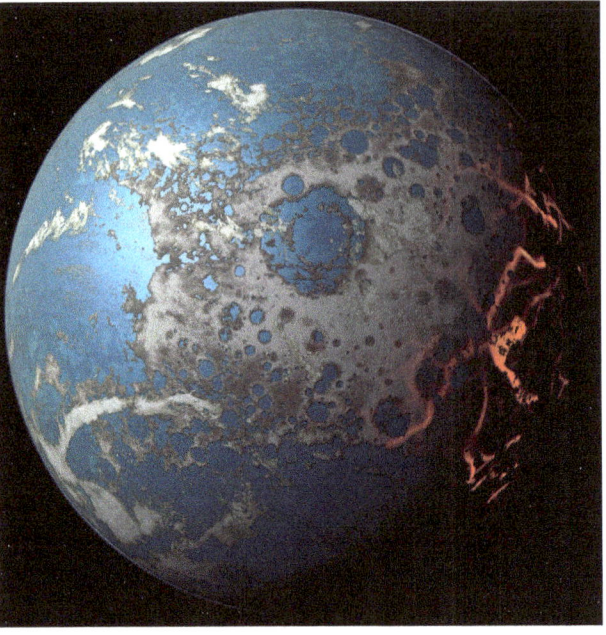

The heavily-bombarded Precambrian Earth.

with no shells, skeletons, etc. Thus, fossils from this eon are rare, and the history of life at the time is not well-documented. Regardless, it is known that **there were numerous types of microbes, aquatic plants, and soft-bodied animals living in the oceans during the Precambrian Eon**.

Organisms present in the Precambrian Eon diversified from single-celled organisms to multi-cellular organisms, aquatic plants, and soft-bodied animals, such as sponges, jellyfishes, and aquatic worms.

Animals without a backbone are called **invertebrates**, and many began making hard parts such as shells, spines, and tough exoskeletons as the Paleozoic Era began. Thus, the number of fossils representing invertebrate life increased dramatically, recording the continuing evolution of life very well. What is seen is an explosion of new life-forms at this time, some of which eventually moved onto the land.

Corals, clams, snails, urchins, crabs, and squids are common examples of living invertebrates, and fossils of their ancestors, and those of many other types of invertebrates, are common in Paleozoic-age marine rocks.

The ancestors of modern plants made the transition from life in water to life on land during the Paleozoic Era, about 500 million years ago. Conditions were quite different back then, and plants formed vast swamps in many areas.

129

Plants began living on land during the Paleozoic Era, about 500mya, and formed vast swamps after that. Later, a variety of **bugs became the first animals to live on land**, the oldest fossil of one being about 425 million years old.

Back in the oceans, the first fishes, and thus the first vertebrates, appeared about 515mya and diversified rapidly. Then, almost 150 million years later, the first amphibians appeared, being the first vertebrates to live on land. About 50 million years after that, the first reptiles appeared, as well. So, the **fishes, amphibians, and reptiles all got their start during the Paleozoic Era**.

Then, catastrophe struck. **At the end of the Paleozoic Era, the worst known mass-extinction in the Earth's history occurred**, which wiped out approximately 95% of all species in the oceans and 70% of land-living vertebrates. So, everything that is still here today evolved from the relative handful of organisms that survived. The cause of this extinction event is not known for certain, but there is significant evidence that it was due to rampant volcanic activity, a large asteroid impact, or both.

Bugs of various sorts were the first animals to live on land.

Fossils show that fishes, amphibians, and reptiles all arose during the Paleozoic Era.

Life began to recover from the extinction event as the Mesozoic Era began, and the first dinosaurs appeared about 230mya. They evolved from reptiles that survived the end of the Paleozoic Era, and went on to become the dominant type of animal on land. Numerous other reptiles took up life in the oceans (ichthyosaurs, plesiosaurs, mosasaurs, and sea turtles) and in the skies (pterosaurs), as well.

Dinosaurs of all kinds, large and small, arose and flourished during the Mesozoic Era.

Several types of marine reptiles and flying reptiles also appeared in the Mesozoic Era. These were not dinosaurs in the water or air though, as they independently evolved from reptiles.

The first mammals appeared during the Mesozoic Era too, which also evolved from reptiles about 195 million years ago. And, the first birds evolved from relatively small dinosaurs, about 150mya. So, the **dinosaurs, marine reptiles, flying reptiles, mammals, and birds all got their start during the Mesozoic Era.**

Mammals evolved from reptiles, and birds evolved from dinosaurs during the Mesozoic Era.

And then, catastrophe struck again. **At the end of the Mesozoic Era, another mass-extinction occurred**, which wiped out as much as 75% of life on land and in the sea, including all of the dinosaurs, marine reptiles (with the exception of sea turtles), and flying reptiles. Many plants and invertebrates also died out, as well as almost all animals that weighed more than about 50 pounds. In this case, it has been well-established that **an asteroid impact was the cause**, and the crater it made has been found.

So, once again, everything that is still here today evolved from the organisms that did survive. With the dinosaurs gone, mammals rapidly diversified and became the dominant type of land animals in the Cenozoic Era. And, some mammals, like seals and whales, took to the sea, as the marine reptiles were gone.

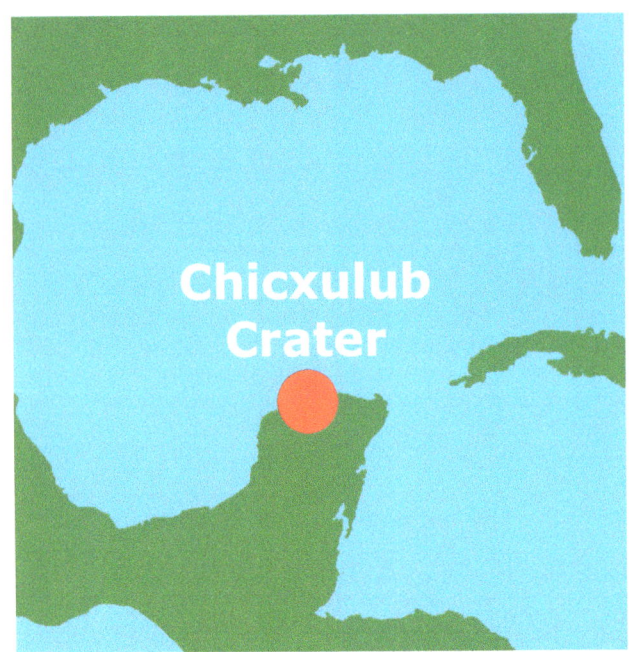 

It cannot be seen at the surface, as it has been filled in over time, but the crater made by the asteroid that caused the end-Mesozoic extinction has been found in Mexico and is about 110 miles in diameter. Laying a circle of this size over Florida provides a scale of reference as to just how large this impact was.

Birds began to flourish in the absence of the flying reptiles, as well. Eventually **hominins**, modern humans and all of their close ancestors, also appeared, with fossil and archeological evidence showing that **the first upright-walking and tool-using hominins arose about 3 million years ago**. Still, modern humans did not arise for another 2.8 million years. So, **our species has existed for only 200,000 years.** We are, of course, still here and the most dominant animal that has ever lived on the planet.

A skull of *Homo habilis*, the first upright-walking and tool-using hominin, which first appeared about 3 million years ago. And, modern humans, *Homo sapiens*, which appeared in Africa about 200,000 years ago, spread across the planet, and genetically adapted to the different environments they moved into.

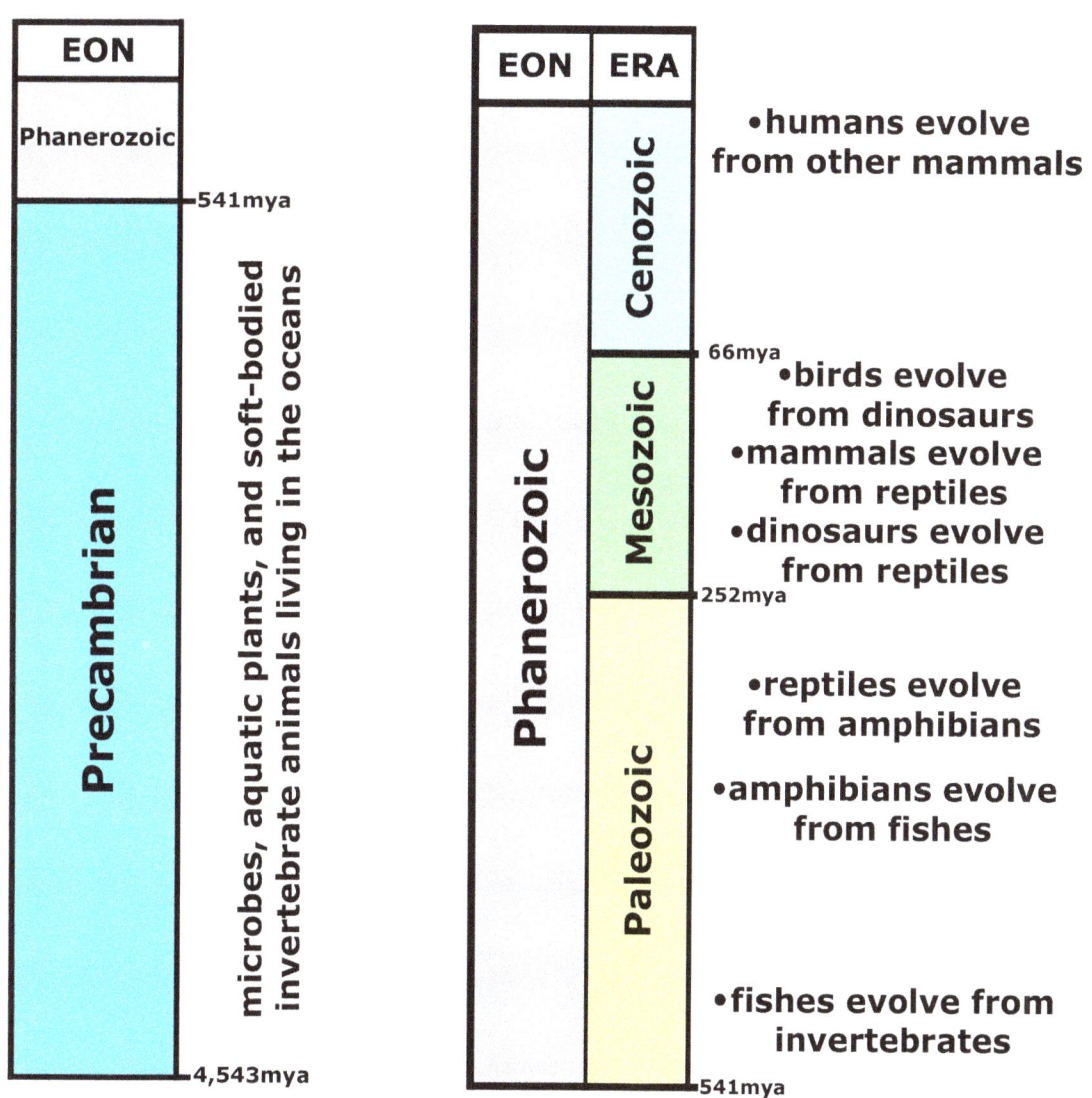

A summary of the types of organisms present at different times, and their evolutionary sequence.

Sample Exam Questions

Ten sample exam questions from each unit of the course are given below. These are typical questions about the material presented in each unit, each of which may or may not be on a given exam. Being able to correctly answer all of these from memory after studying will be a good indication that you are adequately prepared to pass the corresponding unit exam. If you cannot answer these from memory, you are not adequately prepared to perform well on the corresponding exam, and should continue studying.

Unit 1

1) What percent of the Earth's atmosphere is oxygen?

2) What is airflow associated with a Northern Hemisphere anticyclone like?

3) The Subtropical High exists at about what latitude?

4) What are clouds that exist as mid-altitude thin sheets/layers called?

5) What are devices used to measure humidity called?

6) Why don't tropical cyclones form in the eastern part of the South Pacific Ocean?

7) What are lines connecting areas of equal pressure on a weather map called?

8) What scale is used to rank tornados?

9) What type of tropical cyclonic weather system has wind speeds ranging from 39 to 73mph?

10) If a parcel of air has a specific humidity of 5g/kg and a water vapor capacity of 20g/kg, what is its relative humidity?

Unit 2

1) The movement of soil, sediment, or rock by wind, water, or ice is called?

2) Cooling lava with bubbles trapped in it produces a rock with which igneous texture?

3) What is a hot spring that periodically ejects a fountain of boiling water and steam into the air called?

4) What is a large deposit of alluvium found at the base of many mountains called?

5) What is formed when a glacial trough is flooded by seawater?

6) Streams typically form what type of drainage pattern in headlands?

7) What is the process of turning sediments into a sedimentary rock called?

8) What is the inclined layering of sand within a dune called?

9) What type of dripstone can form on the ceiling of a cavern?

10) What is the most abundant element in the Earth's crust?

Unit 3

1) What are the crust and rigid uppermost portion of the mantle collectively called?

2) What scale is used to rank the intensity of an earthquake?

3) What is the Earth's inner core primarily composed of?

4) What type of volcano is Mt. St. Helens?

5) What type of tectonic boundary is found at the Mid-Atlantic Ridge?

6) What type of pluton is concordant and tabular in form?

7) Typically, how thick is continental crust, in miles?

8) What is the name of the "super continent" proposed in the Theory of Continental Drift?

9) What is it called when moist soil/sediment is turned to mud during an earthquake?

10) What type of fault is created when a hanging wall block moves down relative to a foot wall block?

Unit 4

1) The worst known mass extinction in the Earth's history occurred at the end of which era?

2) What is the western part of the North Atlantic Gyre, which runs up the East Coast of the U.S., called?

3) What structures can be constructed along shorelines in order to create or build up a beach?

4) What is a flat-topped seamount called?

5) Surface currents in the oceans are driven primarily by what?

6) In what trench is the deepest place in the world's oceans found?

7) Which part of a passive continental margin is comprised of numerous abyssal fans?

8) The dinosaurs went extinct how many years ago?

9) What's it called when a location experiences the greatest difference between high and low tides?

10) What type of unconformity is found when sedimentary rock is in direct contact with igneous or metamorphic rock?

Answers to Sample Exam Questions

Unit 1

1) 21%

2) air flows down and out, in a clockwise manner

3) 30° north and south

4) altostratus

5) hygrometers

6) the water there is too cool

7) isobars

8) the Enhanced Fujita Scale

9) a tropical storm

10) (5/20)100 = 25%

Unit 2

1) erosion

2) a vesicular texture

3) a geyser

4) an alluvial fan

5) a fiord

6) a dendritic pattern

7) lithification

8) cross-bedding

9) a stalactite

10) oxygen

Unit 3

1) the lithosphere

2) the Modified Mercalli Scale

3) iron and nickel

4) a stratovolcano or composite volcano

5) an oceanic divergent boundary

6) a sill

7) 25 to 40 miles

8) Pangea

9) liquefaction

10) a normal fault

Unit 4

1) the Paleozoic Era

2) the Gulf Stream

3) groins

4) a guyot

5) winds

6) the Mariana Trench

7) the rise

8) 66 million

9) a spring tide

10) a nonconformity

Image Credits

Cover images: National Air and Space Administration; **Pg. 5:** spiral galaxy and Eagle Nebula: NASA; **Pg. 6:** terrestrial planets, Jovian planets, and Pluto: NASA; Earth's spheres: US Geological Survey; **Pg. 9:** seasonal positions: Robin Milkowitz/Greener Pixels; **Pg. 11:** satellite image: Google Earth/Landsat imagery; **Pg. 19:** cloud types: Valentin de Bruyn, Wikimedia Commons, Creative Commons License 3.0; **Pg. 20:** sleet: Runningonbrains, Wikimedia Commons, public domain; hail (L): Public Domain Pictures; hail (R): National Oceanic and Atmospheric Administration; **Pg. 21:** dew: Public Domain Pictures; frost: Public Domain Pictures; **Pg. 22:** air pressure map: modified from National Weather Service; **Pg. 23:** air pressure map: modified from NWS; tropical cyclones: NOAA; **Pg. 24:** jet streams: NOAA; **Pg. 25:** satellite image: NASA; **Pg. 27:** weather map: modified from NWS; **Pg. 28:** satellite image: NASA; radar image: NOAA; weather map: NWS; **Pg. 30:** tornado: NWS; **Pg 31:** satellite image: NASA; **Pg. 32:** cross-section: NASA; typhoon eye: NASA; **Pg. 33:** hurricane season: modified from NOAA; satellite image: NASA; **Pg. 34:** hurricane tracks: NOAA; **Pg. 50:** slump diagram: modified from USGS; slump: USGS; debris flow diagram: modified from USGS; **Pg. 52:** satellite images: Google Earth/Landsat imagery; **Pg. 53:** satellite image: Google Earth/Landsat imagery; **Pg. 54:** satellite images: Google Earth/Landsat imagery; **Pg. 55:** cementation: modified from USGS; **Pg. 56:** aquifer diagram: modified from USGS; **Pg. 57:** geyser diagram: modified from USGS; **Pg. 59:** Winter Park sinkhole: USGS; satellite image: Google Earth/Landsat imagery; **Pg. 60:** glacier diagram: modified from Climatica; **Pg. 64:** glacier diagram: modified from Climatica; **Pg. 65:** satellite images: Google Earth/Landsat imagery; **Pg. 66:** satellite images: Google Earth/Landsat imagery; **Pg. 67:** Sahara Desert: Fiontain, Wikimedia Commons, CCL 3.0; major deserts: modified from USGS; U.S. deserts: modified from USGS; **Pg. 68:** satellite image: Google Earth/Landsat imagery; **Pg. 69:** satellite image: Google Earth/Landsat imagery; **Pg. 71:** sandstorm: US Marine Corps; **Pg. 72:** satellite image: Google Earth/Landsat imagery; **Pg. 73:** cross-section: modified from Robin Milkowitz/Greener Pixels; **Pg. 74:** Pangea: modified from Univesity of Texas Institute for Geophysics; **Pg. 75:** tectonic plates: modified from USGS; **Pg. 76:** crustal ages: Google Earth/US Department of State Geographer; satellite image: Google Earth/Landsat imagery; magnetic reversals: modified from USGS; **Pg. 77:** plate motion: modified from NASA; plate boundaries: modified from USGS; tectonics cross-scetion: USGS; **Pg. 78:** satellite images: Google Earth/Landsat imagery; **Pg. 79:** satellite images: Google Earth/Landsat imagery; San Andreas Fault: USGS; **Pg. 84:** earthquake map: NASA; **Pg. 85:** Ring of Fire: Gringer, Wikimedia Commons, public domain; earthquake hazard map: USGS; **Pg. 86:** tsunami energy: NOAA; **Pg. 87:** lava flow: USGS; **Pg. 88:** underwater pillow lava: NOAA; pillow lava: Avenue, Wikimedia Commons, CCL 3.0; **Pg. 89:** large lava tube: Frank Schulenburg, Wikimedia Commons, CCL 3.0; **Pg. 90:** lava fountain: USGS; satellite image: NASA; volcanic ash: USGS; **Pg. 92:** crater (L): USGS; **Pg. 93:** satellite image: Google Earth/Landsat imagery; fissure eruption: USGS; fumarole (R): USGS; **Pg. 94:** satellite image: Google Earth/Landsat imagery; **Pg. 95:** satellite images: Google Earth/Landsat imagery; **Pg. 96:** satellite image: Google Earth/Landsat imagery; Mt. St. Helens: USGS; **Pg. 97:** satellite images: Google Earth/Landsat imagery; **Pg. 98:** satellite images: Google Earth/Landsat imagery; **Pg. 99:** pyroclastic flow: USGS; satellite image: Google Earth/Landsat imagery; **Pg. 100:** satellite image: Google Earth/Landsat imagery; **Pg. 101:** Ship Rock, Bowie Snodgrass, Wikimedia Commons, CCL 2.0; **Pg. 102:** fold mountain: modified from Public Domain Pictures; satellite images: Google Earth/Landsat imagery; **Pg. 104:** satellite image: Google Earth/Landsat imagery; **Pg. 105:** satellite images with volcanoes: Google Earth/Landsat imagery/NOAA; satellite image: Google Earth/Landsat imagery; **Pg. 106:** satellite images: Google Earth/Landsat imagery; **Pg. 107:** satellite images: Google Earth/Landsat imagery; **Pg. 108:** satellite images: Google Earth/Landsat imagery; **Pg. 109:** satellite images: Google Earth/NOAA/U.S. Navy; **Pg. 110:** satellite images: Google Earth/Landsat imagery; crustal ages: NOAA; **Pg. 111:** satellite images: NASA; plankton: NOAA; **Pg. 112:** gyres (T): modified from Avsa, Wikimedia Commons, CCL 3.0; gyres (B): US Army; **Pg. 113:** surface currents: Windyty.com; **Pg. 114:** seawater density: NASA/Goddard Space Flight Center Scientific Visualization Studio, The Blue Marble Next Generation, data courtesy of Reto Stockli (NASA/GSFC) and NASA's Earth Observatory; ocean currents: modified from Avsa, Wikimedia Commons, CCL 3.0; **Pg. 115:** El Niño: NASA; **Pg. 118:** satellite image: Google Earth/Landsat imagery; **Pg. 119:** satellite images: Google Earth/Landsat imagery; **Pg. 121:** satellite images: Google Earth/Landsat imagery; **Pg. 122:** satellite images: Google Earth/Landsat imagery; **Pg. 124:** unconformities: modified from Woudloper, Wikimedia Commons, public domain; **Pg. 125:** angular unconformity: modified from James St. John, Wikimedia Commons, CCL 2.0; correlation: USGS; **Pg. 128:** Precambrian Earth: MarioPro, Wikimedia Commons, CCL 4.0; bacteria: NIAID, Wikimedia Commons, CCL 3.0; microscopic algae: NOAA; **Pg. 130:** salamander: Christian Jansky, Wikimedia Commons, CCL 3.0; dinosaur: FunkMonk, Wikimedia Commons, CCL 3.0; **Pg. 131:** marine and flying reptiles: Dmitry Bogdanov, Wikimedia Commons, CCL 3.0; **Pg. 132:** satellite image: Google Earth/Landsat imagery; *Homo habilis*: Daderot, Wikimedia Commons, CCL 1.0

Creative Commons Licenses:
1.0: https://creativecommons.org/publicdomain/zero/1.0/legalcode
2.0: https://creativecommons.org/licenses/by/2.0/legalcode
2.5: https://creativecommons.org/licenses/by-sa/2.5/legalcode
3.0: https://creativecommons.org/licenses/by-sa/3.0/us/legalcode
4.0: https://creativecommons.org/licenses/by/4.0/legalcode

www.ingramcontent.com/pod-product-compliance
Lightning Source LLC
Chambersburg PA
CBHW061812290426
44110CB00026B/2860